Competition and Chaos

Competition and Chaos

U.S. Telecommunications since the 1996 Telecom Act

ROBERT W. CRANDALL

BROOKINGS INSTITUTION PRESS
Washington, D.C.

1775 Massachusetts Avenue, N.W., Washington, D.C. 20036
www.brookings.edu

Library of Congress Cataloging-in-Publication data

Crandall, Robert W.
Competition and chaos : U.S. telecommunications since the 1996 Telecom Act / Robert W. Crandall.
p. cm.
Includes bibliographical references and index.
Summary: "Examines how the 1996 Telecommunications Act and its antecedents have affected the major telecommunications providers and analyzes the act's effect on economic welfare in the United States"—Provided by publisher.
ISBN 13: 978-0-8157-1617-4 (pbk. : alk. paper)
ISBN 10: 0-8157-1617-6
1. Telecommunication policy—United States. 2. Telecommunication—Deregulation—United States. 3. Competition—Government policy—United States. 4. Telecommunication—Law and legislation—United States. 5. United States. Telecommunications Act of 1996. I. Title.
HE7781.C667 2005
384'.0973—dc22 2005001648

9 8 7 6 5 4 3 2 1

The paper used in this publication meets minimum requirements of the American National Standard for Information Sciences—Permanence of Paper for Printed Library Materials: ANSI Z39.48-1992.

Typeset in Adobe Garamond

Composition by R. Lynn Rivenbark
Macon, Georgia

Printed by R. R. Donnelley
Harrisonburg, Virginia

Contents

Preface

The last five years have not been kind to investors in the high-technology sectors of the economy. After enjoying a meteoric rise between 1998 and early 2000, most high-tech company stocks plummeted, driving the NASDAQ index down by more than 70 percent in just 18 months. Telecommunications stocks played a major role in this high-tech "bubble," rising to unsustainable heights on the back of exaggerated expectations created by the Internet and falling dramatically in the wake of dashed expectations and a number of corporate scandals.

A major change in U.S. regulatory policy coincided with this telecommunications stock bubble. The 1996 Telecommunications Act opened all telecom markets to competition for the first time and placed regulators in the position of facilitating competition rather than presiding over protected monopolies. This book analyzes the impacts of this tectonic shift in regulatory policy on consumers and telecom companies. I conclude that much of the new policy of assisting new, small-scale entrants was a failure, inducing investors to squander billions of dollars while producing little in the way of new services or innovation. In fact, the new entrants substantially depressed productivity growth in the sector.

The major development in telecommunications since 1996 involves high-speed connections to the Internet. These "broadband" connections were not generally available to the mass market when the 1996 act was passed and therefore did not feature prominently in the debate over future telecom policy. Nor had the new national wireless carriers, such as PCS Sprint, Nextel, or T-Mobile, begun to compete aggressively with the two established wireless carriers that had emerged earlier from the FCC's 1980s policy. The diffusion of broadband, particularly over cable television systems, and the rapid growth of "cellular" wireless systems created a competitive environment that could not have been foreseen in early 1996, when the new act was signed into law.

Nine years after the passage of the Telecommunications Act, it is quite clear that the sector is settling down into a competitive struggle among three alternative communications platforms—namely, those of the wireless carriers, the large traditional Bell telephone companies, and the cable television companies. The new high-speed Internet services are driving technology and investment, and the role of voice communications has been dramatically reduced. Traditional voice telephony is migrating to the Internet along with most other services. In this world, the new small-scale entrants attracted into the sector by the 1996 law are failing rapidly. Even the traditional long-distance companies will not be able to survive as independent entities because the prices they can command have declined so rapidly that they have not been able to cover the costs of the very large investments they made in the 1990s. As this book goes to press, the two largest long-distance carriers (AT&T and MCI) are in the process of being swallowed by the two largest Bell companies (Verizon and SBC).

Regulation works very poorly in a market with rapidly changing technology and strong rivalry among a changing group of large players. As a result, my book recommends a major deregulation of telecommunications so that market forces can drive the necessary investments. And I argue that, to the extent Congress wishes to continue the myriad subsidy programs described as "universal service" policy, it should pay for these policies from general revenues, not from taxes on still-evolving telecommunications services.

Acknowledgments

For numerous comments and suggestions, thanks go to James Alleman, Jerry Ellig, Thomas Hazlett, Eli Noam, J. Gregory Sidak, Hal Singer, and Leonard Waverman. I would also like to thank Simone Berkowitz, Jesse Gurman, and Shen Wu for research assistance. The manuscript was edited by Vicky Macintyre and fact checked by Stephen Robblee. The pages were proofed and indexed by, respectively, Inge Lockwood and Sherry Smith.

1 Introduction

Ten years ago, most of the world's telecommunications companies were state-owned monopolies, performing much like the post offices from which they had sprouted in the early decades of the twentieth century. Those in the United States were different: they had never been government-owned, and the *private* national operator, AT&T, was broken up in 1984 to ensure greater competition in long-distance and equipment markets. Twelve years after the breakup of AT&T, Congress passed the 1996 Telecommunications Act, opening all telecommunications to competition and launching a new era in the sector.[1] No one could have guessed how this era would unfold.

As I describe in this book, "competition and chaos" have enveloped the sector as it gropes toward a new order. This is not to say that the U.S. experience is unique, for similar tumult has descended on most of the world's telecom sector over the past six to eight years. It differs, however, in that U.S. regulators were the first to venture into this brave new world of heavily regulated competition.[2] Furthermore, many of the regulators apparently thought that they could steer a steady course in this direction, with limited disruption.[3] As I demonstrate, they not only failed to achieve this objective, but they contributed—and continue to contribute—to the chaos.[4] Congress invited them to manage competition, and they did so with a vengeance. As this book goes to press, there is little indication that regulators have learned any lessons from the past nine years.[5]

Regulation, Deregulation, and Competition

For decades, the Federal Communications Commission (FCC) and state public utility commissions had regulated telecommunications through an uneasy division of responsibilities.[6] The state commissions oversaw intrastate services—local connections and messages that traveled wholly within a state's borders—and the FCC interstate services. It was not until the late 1970s that the courts clarified the role of the FCC in regulating (or deregulating) the terminal equipment used by businesses and households to connect to the network. Subsequently, the FCC grudgingly admitted competitors into interstate long-distance services, but most states steadfastly continued to refuse to allow competition for most intrastate services. Although regulation might be designed to control monopoly pricing, in telecommunications it was used primarily to redistribute income from businesses to residences and from urban areas to rural areas.[7] This redistribution, which has been defended as "universal service" policy but in fact contributes little to the universality of telephone subscription, continues more than twenty-five years after the FCC began allowing competition, twenty years after AT&T was broken up, and nine years after the 1996 act was passed.[8] Moreover, the implicit subsidies built into the regulated rate structure can continue only if regulators *prevent* competition.

Many discussions of recent U.S. and other telecom policies begin with the notion that the sector is in turmoil because of "deregulation." On the contrary, U.S. policy since 1996 has been far from deregulatory. Local retail telephone rates, intrastate long-distance rates, carrier connection rates, and even high-speed business rates are still highly regulated in most states. More important, the "deregulation" introduced by the 1996 act ushered in a complex new set of regulations involving the provision of wholesale services by incumbent local carriers to their new competitors. Since that time, a large number of other countries have followed the United States in erecting similar wholesale-access regulations to promote competition.

In a later chapter, I analyze the effect of the new regulatory regime on competition and prices in local telecommunications in the United States, but for now it is worth noting that telecom liberalization has diverged from earlier policies in the transportation and energy industries. In the 1970s and 1980s, the United States opened the airline, air cargo, trucking, and railroad industries to competition without increasing regulation. Indeed, the commissions regulating these industries were soon abolished, along with most rate regulation. Congress did not find it necessary or prudent to require carriers in these industries to sell their services or lease their facilities to rivals at regulated rates. By contrast, legislators viewed a large part of the distribution network in telecommunications, and later in electricity, as a natural monopoly. As a result, the 1996 act instructed regulators to determine which incumbent-carrier facilities should be made available to entrants and to establish the cost basis for wholesale rates for such facilities, two issues that have tied up the regulators and the courts for most of the past nine years.

The U.S. experiment in regulated competition has had a large impact because most other countries of the Organization for Economic Cooperation and Development (OECD) subsequently followed it.[9] The Europeans, in particular, have adopted a more modest version of mandated network sharing, as have the Japanese and the Canadians. Given its earlier successes in the deregulation of transportation, energy, and financial markets, the United States now serves as an example to most of the world in the "deregulation" of important network industries. This deference to U.S. policy may change in view of the chaos resulting from U.S. telecom and electricity regulation.[10]

The Financial Market "Bubble"

The first four years after the passage of the 1996 act were exhilarating for many participants in the telecommunications sector. Investment soared as stock market valuations rose at remarkable rates. U.S. long-distance companies, wireless companies, and the new local carriers ushered in by the act saw their stock prices rise by 500 percent or more in 1998–2000 as the Internet boom captured the imagination of investors. Capital spending rose to historic levels, fueled by expectations that the Internet would require enormous increases in telecom capacity. New local companies raced to take advantage of the stock market euphoria, and wireless companies bid aggressively for new spectrum auctioned by the FCC. A number of major companies were started in the late 1990s or formed from a series of mergers, including WorldCom, Qwest, and Global Crossing.

By the middle of 2000, it was apparent that the very large rise in stock market values in Internet-related companies, including telecommunications carriers, could not be sustained. The U.S. NASDAQ average, dominated by technology stocks, began to fall as dramatically as it had risen, declining from more than 5,000 in March 2000 to about 1,114 in October 2002, roughly equal to its level in February 1996.[11] The new local telecommunications start-ups fell by much more, and most were forced into bankruptcy. The long-distance companies met a similar fate, exacerbated by various accounting scandals arising from their desperate attempt to maintain the illusion of growth. Because many of these telecommunications carriers had expanded with loans from their equipment suppliers, the latter, such as Lucent and Nortel, were also devastated.

The United States was not alone in experiencing this boom-bust cycle. The value of telecommunications equities soared throughout the developed world, propelled by excessive enthusiasm for the Internet and the new technologies.[12] The prices paid for spectrum in European auctions to accommodate the "third generation" (3G) of wireless services reached historic levels in 2000.[13] Newly privatized carriers, such as British Telecom and France Telecom, expanded aggressively beyond their national borders through acquisitions and new capital spending.[14] When the equities market soured on these and other Internet-related stocks, many of the major telecom companies scrambled to scale back their operations, while large numbers of new entrants collapsed. As capital spending plummeted, the world's telecom sector sank into a three-year depression, from which it is only now slowly recovering.

Chaos, Regulation, and Market Structure

No regulator faced with the new requirements of the 1996 Telecom Act could have foreseen this turmoil. Looking at the sharp rise in equity prices in 1998–2000, regulators must have thought they were on the right track at first. After all, the skyrocketing value of the new entrants' stocks was attracting large amounts of capital, and the prices of the local incumbents' stocks had not fallen measurably below overall market averages. Surely, these were signals that the burgeoning competition was going to provide a bonanza of benefits.

Once the Internet and telecom stocks went into a dive, everything changed, including faith in the new regulatory regime. On the way down, new local entrants complained bitterly that the regulated wholesale rates were too onerous to let them use the incumbents' networks, and that the regulatory rules were not always enforced. The resulting accounting scandals left two of the four largest long-distance companies in a state of collapse and forced a third to dismiss its senior officers.[15]

In 2000–02 a new set of issues appeared. The Internet was creating a demand for higher-speed, "broadband" connections in residences and small businesses that the incumbent local telecoms were unable to satisfy without substantial new investment in network facilities. But would these companies invest in such facilities if the entrants, now under enormous duress, could immediately lease them at low, regulated rates? For their part, the struggling entrants were asking for even more favorable terms for access to the incumbents' facilities. How was a regulator to respond to these pressures? In past decades, a regulator's only worry was to allow the monopolist a sufficient return on capital to stay in business and to maintain service quality. Now the issue was how new technologies could be deployed to provide innovative services when promoting investment in these technologies was likely to conflict with keeping the new entrants alive.

By 2001 it was also clear that telecom revenues were no longer increasing, particularly among the traditional wire-based carriers. As the prices for using the network dropped in response to competition, total revenues fell because the Internet was not increasing network use fast enough to offset the effects of declining prices. Wireless or "cellular" carriers were rapidly diverting traffic from the wire-based carriers just as the new local entrants were beginning to lure large numbers of subscribers away from the incumbent local companies. Ebb tide was not the best time to launch more boats.

Yet another issue concerned the incumbent Bell carriers, which had been kept out of long-distance services since being divested from AT&T in 1984. The 1996 act would allow them back once they proved that they were cooperating with regulators in providing entrants with access to their networks. For the first four years, regulators were reluctant to approve Bell entry into long distance. As the entrants raced ahead to enroll 10 percent or more of the country's subscriber lines in 2001–02, the regulators could no longer deny the Bell companies the right to offer long-distance services. In one state after another, these companies added to the downward pressure on long-distance rates initiated by the wireless carriers in 1999. The troubled long-distance companies—WorldCom, Qwest, Global Crossing, and even AT&T—now operated under a new set of competitive pressures. As this book is being written, it is far from clear that any of the independent long-distance companies can survive.

The Future?

As 2005 dawns, few observers, regulators, or industry participants can be sure how the U.S. or world telecom sector will evolve. Past exercises in true

U.S. deregulation have often been followed by years of turmoil.[16] The airline industry, for example, is going through another series of bankruptcies and near bankruptcies twenty-seven years after deregulation began because newer U.S. carriers, such as Southwest, AirTran, Jet Blue, and Frontier, are now wreaking havoc on the older industry titans. Similarly, the U.S. railroads have passed through a long period of consolidation and turmoil in the twenty-five years since they were more or less deregulated. Clouds of uncertainty also surround recent attempts to open the U.S. electricity sector to competition. Deregulation or market liberalization in these sectors, however, does not face the same profound technological revolution that is occurring in telecommunications. Hence the 1996 act simply exacerbated a turmoil that seemed bound to occur in any event.

In this book, I attempt to look past the regulatory debates of today to the likely evolution of the telecom sector in the next few years. This is a hazardous enterprise under any circumstances, but it is particularly risky in the unsettled regulatory and market conditions of 2005. I try to distinguish the survivors from those who will likely fail, while providing a policy analysis along the way. My conclusion is that "deregulation" requires *deregulation*: that is to say, the telecommunications sector has been overregulated in the past nine years. Despite having the best of intentions, regulators have not facilitated competition but have likely delayed it. Furthermore, they have failed in their attempt to allow new or established companies to survive without their own connections to customers. Forcing the incumbent local companies to offer their facilities at regulated, wholesale prices to entrants is not only a mistake; it is an exercise in futility.

2 Opening Telecom Markets: The 1996 Telecommunications Act

Before 1996 the U.S. telecommunications sector was mired in a policy morass from which there appeared to be no escape. Though never formally shown to be a "natural monopoly," it was heavily regulated by politically responsive officials intent on using their powers for income redistribution.[1] In this pursuit, the regulators often protected the monopoly power of incumbents so as to have monopoly rents to shift among constituent groups.[2]

In the 1950s, however, the courts began to get involved as competitors stormed the walls. By the 1980s court challenges based on the country's antitrust laws were succeeding, and the courts became yet another source of industry regulation that created its own problems. The largest U.S. telephone company, AT&T, was broken up under a 1982 federal antitrust decree that held sway for fourteen years, until the administration of this decree became so cumbersome that Congress was induced to act. The 1996 Telecommunications Act was its way of resolving the disputes that had arisen in enforcing the decree and regulating a balkanized telecommunications sector.

From Antitrust to a New Era

Competition in U.S. telecommunications actually began much earlier, when the courts forced the Federal Communications Commission (FCC) to open the interstate long-distance market in the 1970s.[3] It opened the market for customer equipment of its own volition at about the same time.[4] The effect was a series of attacks on AT&T's entrenched position in these two markets by new carriers and by a variety of equipment manufacturers. This competitive struggle quickly shifted from the marketplace to the courtroom as the principal new long-distance entrant, MCI, and others alleged that AT&T was engaged in anticompetitive practices in denying entrants access to their customer lines, without which they could not survive. These antitrust suits of the early 1970s reached a crescendo when the government filed its stunning Sherman Act suit against AT&T in 1974.[5] After years of legal maneuvering, the case went to trial in 1980. By August 1981, it was clear that the government would win the case at the district court level. As a result, AT&T negotiated a consent agreement in 1982 that required it to divest itself of its local telephone operating companies in 1984.[6] These operating companies, organized into seven regional Bell holding companies, would be barred from manufacturing telephone equipment, from offering long-distance services, and even from developing enhanced information services.[7]

The long-distance and manufacturing line-of-business restrictions remained in place for twelve years, much to the consternation of the divested Bell companies. To begin with, they created an enormous amount of legal wrangling over the complexities in defining and delineating the market boundaries of services the Bell companies could offer under the decree.[8] More important, the decree established a vertically fragmented and inefficient structure for the telecommunications sector, which is now disappearing. No other country has attempted to balkanize its telecommunications sector in the manner of the decree, separating "local" markets from "long-distance" markets.[9] When the Internet began growing rapidly in the mid-1990s, the Bell companies could not participate in this new medium, even though a call to an Internet service provider (ISP) is generally a local call. A local call to an ISP triggers communications that eventually become "long distance," thereby running afoul of the line-of-business prohibition on the divested local Bell companies.

Similar problems occurred in the enforcement of the manufacturing line-of-business restriction. Given the rate of technological change in electronics and communications in the 1980s and 1990s, one would have expected the major telecom carriers to engage in a variety of joint ventures with equipment suppliers to develop, test, and implement new technologies. However, the decree prohibited the divested Bell companies from entering such endeavors. The decree was more lenient on joint ventures involving software, but the dividing line between software and hardware was changing and created substantial room for disputes that had to be taken to the court enforcing the decree.[10]

After more than a decade of legal wrangling, the Bell companies and their adversaries sought relief from Congress, which struck a compromise between the Bell companies and the long-distance carriers, including AT&T, MCI-WorldCom, and Sprint. The new law would allow the Bell companies to enter long-distance markets if the long-distance companies and other firms could enter the local-access markets historically dominated by the Bell companies and other incumbent local exchange carriers (ILECs), such as GTE, United Telephone, Frontier, and Central Telephone. The new law not only opened *all* telecommunications markets to competition but also provided a maze of new requirements for the FCC and state commissions to implement or for the affected parties to contest in court. These provisions substituted for the AT&T consent decree, which was abolished by the 1996 act.

"Opening" Local Markets

The 1996 Telecommunications Act lays out its prime purpose in its most straightforward provision: "No State or local statute or regulation, or other State or local legal requirement, may prohibit or have the effect of prohibiting the ability of any entity to provide any interstate or intrastate telecommunications service."[11] Had the act stopped there and ended the 1982 AT&T consent decree, much of the legal turmoil that followed (and continues) might have been avoided. Unfortunately, the act contained more than 100 additional pages of detailed instructions to regulators that would unleash years of disputes, legal appeals, and general uncertainty, many of which remain unresolved.[12]

These detailed provisions consist largely of prescriptions for opening local telecommunications markets to competition. For example, they require the incumbent local companies, principally the Bell companies, to interconnect with the entrants at any "feasible" point, to lease their "unbundled" network facilities to entrants at some measure of cost to be determined by the regulators, and to offer their telecommunications services for resale at wholesale discounts reflecting the avoided costs of retailing the service. A variety of other provisions relate to poles, ducts, emergency 911 services, numbering and number portability, directory services, and the services required to switch customers to the new entrants.

Many of these regulations had never been implemented before, especially in regard to network unbundling, number portability, or large-scale, real-time transfer of customers from an incumbent telephone company's network to those operated by new entrants. Indeed, there was virtually no experience with local competition for residential and small-business subscribers anywhere in the world.[13] Nevertheless, the Bell companies and other incumbents were required to adhere to these untested rules, and the Bell companies' compliance on a state-by-state basis was a condition for gaining entry into long-distance services originating in each state.

Network Unbundling

In probably its most controversial provision, the 1996 act required incumbent carriers to provide entrants with access to their unbundled facilities at regulated rates: "[The incumbent carrier has t]he duty to provide to any requesting telecommunications carrier for the provision of a telecommunications service nondiscriminatory access to network elements on an unbundled basis at any technically feasible point on rates, terms, and conditions that are just, reasonable, and nondiscriminatory."[14] The FCC was to identify which facilities had to be unbundled: "In determining what network elements should be made available . . . , the Commission shall consider *at a minimum* whether—[A] Access to such network elements as are proprietary in nature is necessary; and [B] The failure to provide access to such network elements would impair the ability of the telecommunications carrier seeking access to provide the services that it seeks to offer"[15] (emphasis added).

The FCC took this sweeping language as an invitation to require the unbundling of virtually everything in the incumbent carriers' existing networks.[16] Even after the Supreme Court remanded this approach for the FCC's reconsideration, the commission was by and large unmoved.[17] Under its revised rules, the FCC mandated that entrants should still be able to obtain virtually all of their facilities at wholesale regulated rates from their rivals, the incumbents.[18]

The economic justification for requiring incumbents to share their facilities with entrants is that some capital facilities cannot be replicated economically by entrants. The economies of density or "fill" may preclude the duplication of network facilities such as conduits, pole lines, and wires. But the act failed to refer to "essential facilities" or to "natural monopoly." Rather, it suggested that anything that entrants needed to fulfill their business plans should be made available if the absence of such wholesale offerings "impaired" their ability to compete.

Hence the FCC dictated that the "local loops," the wires connecting the subscribers' premises to the telephone companies' switching facilities, be unbundled and made available to competitors. Central office switching, the transport facilities between the companies' offices, numbering resources, network intelligence, and operating and support systems also had to be unbundled so that an entrant lacking the wherewithal to build its own facilities could compete. Furthermore, incumbents had to *share* the individual copper loops that terminate in a subscriber's premises, thereby allowing an entrant to lease the upper frequencies on the loop so as to be able to provide high-speed, broadband Internet service without having to provide ordinary telephone service. This "line-sharing" requirement came a few years after the commission's initial 1996 unbundling rules, but it was overturned by the U.S. Court of Appeals in 2002 in a decision that also asked the FCC to reconsider its approach to unbundling once more.[19]

The 2002 reversal sent the entire unbundling regime back to the commission for a second time. As a result, seven years after the passage of the 1996 act, one of its most important provisions was still in dispute. In its 2003 response, the FCC failed to offer a means of resolving the disputes over network unbundling.[20] Rather than clearly identifying the network facilities that must be unbundled, its 2003 rules left the final decisions to state regulators. After first signaling its decision in February 2003, the commission waited until August to publish 600 pages of new rules containing a large number of ambiguities. For instance, it had announced in February that it would no longer require the incumbents to share their copper loops with entrants, thereby denying the entrant the opportunity to use only the upper frequencies required to deliver new "broadband" services.[21] In the final rule, it allowed entrants to continue line sharing to serve their existing customers for an indefinite period and required the incumbents to offer "line splitting" on lines that they lease to competitive local exchange carriers (CLECs), thereby enabling these lessees to share their lines with third-party suppliers of broadband services.

Given the ambiguities in this newest version of the unbundling rules and the decision to leave important decisions to the states, they were appealed to the federal courts once again. In 2004 the D.C. Circuit, in a stinging rebuke to the commission, overturned the latest FCC attempt to establish the dimensions of network unbundling, but it left intact the decision to end line sharing.

In December 2004, the FCC was in the process of rewriting the rules once again, but it is now clear that the degree of network unbundling will be sharply reduced, particularly the requirement that the incumbent local carriers offer the entire complement of network facilities as an "unbundled network element platform" (UNE-P). As I discuss in chapter 5, this platform has generally been made available to entrants at discounts of 50 to 60 percent from average retail revenues per line, thereby providing the entrants with a substantial prospective margin and greatly reducing the incumbents' revenues per line. This regulatory arbitrage opportunity is now apparently coming to an end.

Terminating Calls

Competition could not work in a network industry if networks did not interconnect. If new entrants were forced to negotiate such interconnection with incumbents that had virtually all of the subscribers, the entrants would be at a substantial disadvantage. For this reason, the act requires that all interconnection rates for connecting traffic be the same for entrants and incumbents. If the incumbent were to insist on, say, $0.10 per minute for terminating calls that originate on the entrants' networks, the entrants must also be paid $0.10 per minute for terminating calls from the incumbents' network. This requirement for "reciprocal compensation" reduces the incentive for the incumbent to exploit its market power in terminating calls.[22] Indeed, given the ability of entrants to attract customers with a large share of terminating traffic, this requirement succeeds in keeping the rates for interconnecting local networks reasonably low.[23]

Mandatory Resale

In addition to offering new local entrants access to unbundled network elements, the incumbent telephone companies must allow the entrants to market, or "resell," their services. For these services entrants must pay *regulated* wholesale rates that are equal to the retail rate less the incumbent's "avoidable cost" of retailing the service, namely, the marketing, billing, and collection costs. Because incumbents have such modest retailing costs, the discounts offered to entrants under this provision have been generally about 15 percent, a level that is simply not very attractive to entrants.[24]

Collocation

Since incumbent local carriers were required to interconnect at any feasible point in their networks and to offer their network piece-parts as unbundled network elements, they also had to allow entrants to "collocate" their equipment in the incumbents' premises at "just" and "reasonable" rates. Had this requirement not been included in the act, interconnection and unbundling might only have been interesting *curiosa,* because entrants would have had difficulty connecting to the incumbents' networks at various nodal points. Nevertheless, this provision created substantial opportunities for controversy and contentious negotiations because incumbents had not designed their facilities to accept rivals' equipment.

Access to Rights of Way

Even before the 1996 act, regulated telecom carriers were required to allow their rivals access to conduits, poles, ducts, and rights of way on reasonable terms. This requirement is now incorporated into the general rules for interconnection under the 1996 act, the purpose being to ease entry and assist new entrants.

Initially, the FCC interpreted the law's various requirements in a manner that would provide the greatest possible boost to entrants.[25] This experiment is now coming to a conclusion as entrants fail and regulators are forced to roll back many of the more ambitious attempts to promote entry (see chapter 4).

Opening the Long-Distance Market

It is often thought that although the 1996 act opened local telecommunications to competition, no new legislation was required to create competitive long-distance markets because they were already highly competitive. The interstate long-distance market had indeed been open to entry since the FCC lost its battle in the federal courts in 1977 over its right to exclude competition.[26] Moreover, the 1982 AT&T decree required the divested Bell companies to provide "equal access" to all long-distance competitors. Despite these pro-competitive policies, the long-distance market remained heavily concentrated through 1995, at which time the largest three long-distance carriers accounted for 75 percent of revenues. The Herfindahl-Hirschman index of concentration was in excess of 3,000, far above the Department of Justice's threshold of 1,800 that defines a "concentrated" market.[27] The average interstate and international revenue per minute was still 17 cents, even though interstate access charges had been reduced to about 6.5 cents per minute by 1995.[28] As this margin over access charges clearly reflected, the market was not fully competitive.

Given the available long-distance margins, the three major carriers spent billions of dollars a year trying to lure customers from one another.[29] They also worked tirelessly to keep the Bell companies excluded from interLATA long-distance services, an exclusion that continued after 1996.[30] Under the 1996 act, in order to be allowed to offer long-distance services across LATA boundaries from a given state, a Bell company would have to satisfy the competitive "checklist" of market-opening conditions and convince first the state regulatory commission, then the Justice Department and the FCC, that it had indeed complied.[31] For three and one-half years, no Bell company was able to gain such permission in any state. After five years, the Bell companies had gained access in only four states (see chapter 5).

This slow progress can be attributed to the opponents to Bell entry, the long-distance companies and competitive local carriers, who could plausibly argue that the Bell companies had not done "enough" to facilitate entry into their local markets. Since there was no experience in such "market opening," there were no benchmarks, but simply endless arguments before the three regulatory bodies required to approve each application. This feature of the 1996 act is now little more than a historical footnote, for by December 2003 the Bell companies had been able to persuade regulators that they had satisfied the act's checklist in all of the forty-eight mainland states and the District of Columbia. With this entry and the competition from wireless services, long-distance rates have declined dramatically. The independent long-distance company is now an endangered species, created by litigation, perpetuated by regulation, and then dashed on the coals of the fires of competition (see chapter 6).

Wireless Communications

Not too long ago, wireless rates were so high that few persons would choose to own a cell phone. Today there are about 170 million wireless subscribers, or more than one cell phone for every two men, women, and children in the United States.[32] Within the next year, the country is likely to have more wireless telephones than wired ones.

The explosion of wireless telephony began at roughly the same time as the experiment with local entry into wired telecommunications in the United States. The major difference in these two revolutions is that the wireless one was essentially unregulated whereas local wireline competition was heavily regulated, as described in the preceding paragraphs. Under the 1993 Omnibus Budget Reconciliation Act, the FCC was instructed to launch this competition by auctioning new spectrum. Federal and state regulators were not supposed to regulate wireless rates unless it could be shown that wireless carriers had a "dominant" market position (for details, see chapters 3 and 7). Today five major, national wireless carriers compete aggressively with one another and with wire-based telecommunications companies, and wireless rates go unregulated.[33]

The 1996 act had little direct effect on wireless carriers, except in the area of interconnection charges. Setting the prices paid to terminate calls on wireless handsets is a controversial regulatory issue in most countries.[34] Unlike the United States, most countries do not allow wireless carriers to charge their subscribers for incoming calls. Instead, the carriers are paid for such calls through access charges levied on the carrier through whose network the call originates. In the United States, wireless subscribers are charged for both inbound and outbound calls. Under the 1996 act, all such local calls are subject to the reciprocal compensation rule, meaning that wireless carriers are compensated for terminating inbound calls at the same rate as are local wire-based carriers, in addition to the calling charges paid directly by the wireless subscriber for incoming calls. The reciprocal compensation component of these charges is generally less than 1 cent per minute, and the total charges borne by the subscribers are generally much lower in the United States than in other countries because of the aggressive competition over both inbound and outbound rates.[35]

Cable Television

Cable television has traditionally been regulated under a different title of the Telecommunications Act than the one that applies to telecom's common carriers. Legislation passed in 1992 to regulate cable television rates did not apply to services delivered by cable systems that were not traditional video services. The 1996 act rolled back much of the cable television regulation established in 1992 but did not address the regulation of the other cable services, such as the new "broadband" Internet (cable modem) services or telephone services provided over cable television networks.[36] These latter regulatory issues currently remain in limbo.[37]

The uncertainty surrounding the regulation of broadband services or ordinary voice-grade telephone services offered by cable companies is a byproduct of technological change. Cable television never fit easily into the traditional regulatory framework that divided carriers into "broadcasting," "telecommunications" (generally, telephone services), and "information services." As cable companies evolve into one of the most important competitors for *all* communications services—voice, video, and data—it will be necessary to eliminate these uncertainties. One such question is whether cable modem services fall under "information" or "communications." Another is whether voice telephony offered over the Internet, as cable companies are likely to provide it, should be subject to the same intercarrier compensation payments or access charges as traditional telephone services. These questions were left unanswered by the 1996 act, and the FCC's attempt to answer them has been delayed by a court reversal.[38]

Conclusion

The 1996 Telecommunications Act has certainly changed the telecommunications landscape in the United States. Entry into all telecom services has been liberalized, and local telephone companies are faced with competition for the first time in nearly a century. Unlike earlier exercises in market opening, however, the 1996 act did not advance "deregulation." Instead, it established a major new set of wholesale regulations designed to facilitate the entry of new carriers. This new regulatory regime has been extremely controversial, resulting in one legal appeal after another. As the following chapters demonstrate, it is far from clear that these new regulatory rules have advanced sustainable entry into telecommunications.

Fortunately, the 1996 act did not extend a new regulatory regime into the domain of wireless telecommunications or cable television. On the contrary, it reduced cable television (video programming) regulation while continuing a policy of largely unregulated competition for wireless services. These cable and wireless companies, not the new entrants, are likely to provide the most potent competition for the established telephone companies.

3 The First Eight Years under the New Law

The 1996 Telecommunications Act not only freed the Bell operating companies from the 1982 line-of-business restrictions but also exposed them to extensive local competition for the first time. Long-distance companies could now enter local markets and therefore offer integrated service packages to their customers. With Internet use also gathering momentum and equity markets enjoying a surge of historical proportions, the United States found itself in the middle of a seemingly unending economic boom.

The Growth of Telecom: Rhetoric versus Reality

A hot topic of this period was the potential of the "information superhighway."[1] The declining cost of fiber-optics transmission and the progress driving microprocessor technology (at a rate commensurate with Moore's Law) led some to predict that communications bandwidth would soon be virtually free.[2] Households, businesses, and every profession from physicians to teachers would be able to send and receive high-speed video images that would substitute for personal diagnoses, provide remote monitoring, allow remote tutoring, and supply personalized access to entertainment. Once the telecommunications sector was opened to competition and deregulated, innovation could flourish, thereby allowing subscribers access to new services and providing existing services at dramatically lower prices.

The Investment Boom

Unfortunately, as explained earlier, the 1996 act did not deregulate telecommunications. Instead it created a vast new system of wholesale-price regulation that was only vaguely spelled out in the statute. Incumbents' local telecommunications services, in particular, would continue to be intensely regulated by the states and the Federal Communications Commission (FCC). By contrast, long-distance and wireless services had been more or less deregulated before the act was passed, so both sectors had already begun to invest heavily in infrastructure in the early 1990s. The lure of the Internet would entice them to accelerate this investment after 1996. In addition, new local carriers sprouted from everywhere and were able to attract enormous amounts of capital. The result was an investment boom that continued for more than five years.

Between 1987 and 1996, nominal capital spending increased at an average rate of 4.8 percent a year (figure 3-1).[3] In the next four years, it soared to more than 20 percent a year. By 2000 real capital spending had risen 148 percent from its 1996 level. This surge was accompanied by an even greater rise in the price of telecom equities. In 2000–01 this stock market "bubble" burst (figure 3-2).[4] The new local telephone carriers, the competitive local exchange carriers (CLECs), suffered the greatest decline, followed by the wireless carriers and the long-distance companies.[5] Many of the CLECs and long-distance companies were forced into bankruptcy, and there is now very little market value left in these two segments of the telecom sector.

Between January 2000 and January 2003, the value of most U.S. telecommunications carriers, particularly the long-distance companies and new local entrants, dropped precipitously, their total loss reaching almost $1 trillion. As a result, capital spending declined substantially in 2001–02 and continued to do so as the bankruptcies among telecommunications carriers reached alarming levels (see table 3-1).

The capital-spending boom spread far beyond fiber-optic transmission facilities.[6] Capital spending by new local carriers increased from virtually nothing to more than $20 billion in 2000.[7] In the wireless sector outlays jumped from $8.5 billion in 1996 to $18.4 billion in 2000, and among the regional Bell companies (including GTE) capital spending rose from $20.8 billion in 1996 to $35.7 billion in 2000, even though these firms were by and large banned from interstate communications.[8]

Once the stock market bubble burst and scores of carriers collapsed, a climate of depression descended on the U.S. and world telecommunica-

Figure 3-1. *U.S. Telecommunications Sector Capital Expenditures, 1987–2003*

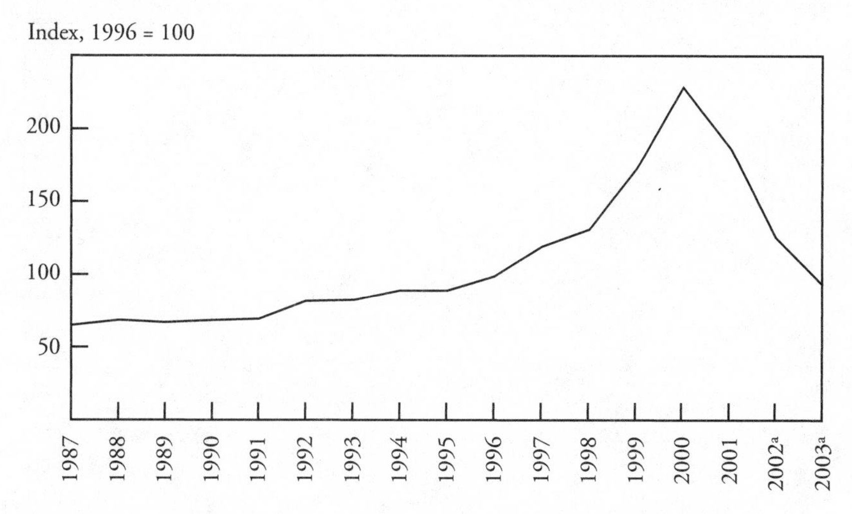

Source: U.S. Department of Commerce, Bureau of Economic Analysis ["BEA"](2003).
a. Estimated by the author based on current dollars.

tions sectors, intensified by the excess capacity in data and voice transmission the boom had created. The long-distance and local failures, combined with the reductions in Bell-company cash flow, brought total capital spending in the U.S. telecom sector down to about $50 billion in 2003, which was even less than its 1996 level. Telecom stock prices experienced a similar collapse in Europe and, to a lesser extent, in Asia (see chapter 9). In just two years, the unbounded optimism shifted to gloom and doom. As 2004 dawned, much of the telecom sector remained depressed—the only exception being the wireless sector, which was buoyed by a battle between Vodafone and Cingular to buy AT&T Wireless.[9]

Output and Revenue Growth

Surprisingly, telecom output and revenues did not accelerate very much in the late 1990s, despite the surge in investment, Internet explosion, and strong economic growth. Between 1987 and 1995, the growth in nominal telecom output ("gross domestic product") and in capital spending was virtually identical.[10] After 1995 the industry's output (as measured by gross product) accelerated only slightly in nominal dollars, rising from a 4.6 percent annual growth rate during 1987–95 to just 6 percent from 1995 to

Figure 3-2. *Telecom Company Stock Prices, 1996–2004*

Index, February 1996 = 100

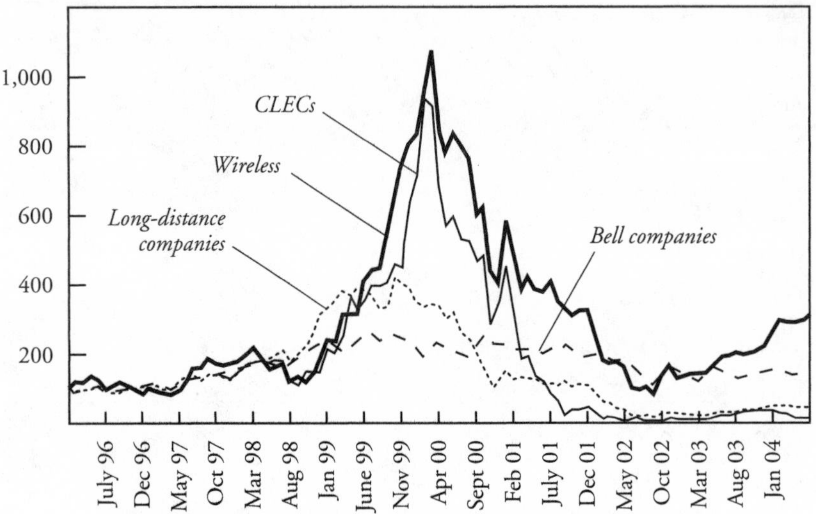

Source: Author's calculations from closing prices on www.finance.yahoo.com.

2000, investment increased by nearly 19 percent per year after 1995 (figure 3-3). The growth in telecom output was unexpectedly modest, given that real gross domestic product (GDP) grew at an average annual rate of 4 percent during this period; it was also far below the growth in nominal capital spending.[11] Indeed, the value of telecommunications output in nominal dollars rose more slowly than did durable-goods manufacturing in the late 1990s.[12]

Virtually all of the recent growth in telecommunications services output and revenues has come from the wireless sector. As figure 3-4 shows, total revenues of the local exchange companies and long-distance carriers have been static since 1996. End-user *wireline* revenues even decreased by about $2 billion, dropping from $151.5 billion in 1996 to $149.2 billion in 2003, despite the phenomenal growth of the Internet.[13] In real inflation-adjusted terms, these figures indicate a decline of about 2 percent a year.[14] This was surely not the explosive growth anticipated from the information technology (IT) revolution and "deregulation."

The lack of revenue growth does not mean that there was no *real* output growth in telecommunications—which was actually substantial,

Table 3-1. *Recent Bankruptcies of Telecommunications Carriers*

Long-distance and wholesale fiber carriers	*Competitive local-exchange carriers*	*Wireless carriers*
360 Networks	Adelphia Business Solutions	Advanced Radio Telecom
Aleron	Allegiance Telecom	GlobalStar
Cambrian Communications	Broadband Office	Iridium
Carrier 1	Convergent Communications	Metricom/Ricochet
Digital Transport	Covad Communications	Motient
Ebone/GTS	CTC Communications	Nextwave Telecom
Enron Broadband	e.spire	OmniSky
FLAG Telecom	FastComm	StarBand
Global Crossing	ICG Communications	Teligent
GST	ITC^DeltaCom	Winstar
Iaxas	Knology	
Impsat	Metromedia Fiber Network	
KPNQwest	McLeod USA	
Neon Communications	Mpower	
Pangea	Network Plus	
Sigma Networks	New Global Telecom	
Sphera	NorthPoint	
Storm Telecommunications	NTL	
Teleglobe	RCN	
Telergy	Rhythms NetConnections	
360networks	RSL	
Velocita	Song Networks	
Viatel	XO Communications	
Williams Communications	WINfirst	
WorldCom	Yipes Communications	
	Zephion	

Source: *Converge! Network Digest*, accessed at www.convergedigest.com/Mergers/financialarticle.asp?ID=4160 (January 20, 2003) and http://quote.yahoo.com.

though not as high as expected.[15] Rather, output did not increase enough to offset the steep decline in prices that was occurring. Indeed, the price of transmitting the trillions of bits of information generated by the Internet fell dramatically, but demand did not rise sufficiently to boost revenues. As is clear from figure 3-5, which shows the recent trend in the telecom industry's contribution to GDP in real and nominal terms, real output in this sector failed to double even every two or three years after 1996.

Figure 3-3. *Current Dollar Investment and Output in the U.S. Telecommunications Sector, 1987–2001*

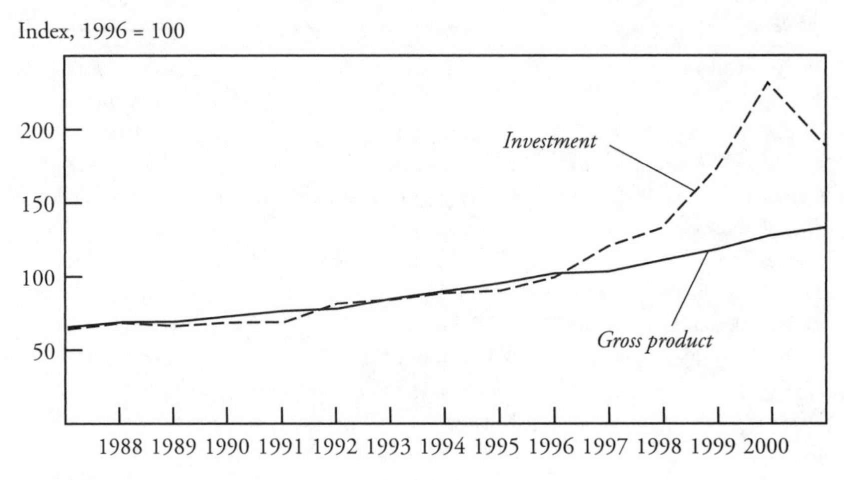

Source: BEA (2003).

Productivity Growth

Perhaps the most striking aspect of the IT revolution of the late 1990s was the *decline* in productivity growth in wired telecommunications services (see figure 3-6). In the nonfarm sector, the average rate of growth of U.S. productivity exploded in 1996–2000, rising by more than two-thirds

Figure 3-4. *Total End-User Telecom Revenues, 1996–2003*

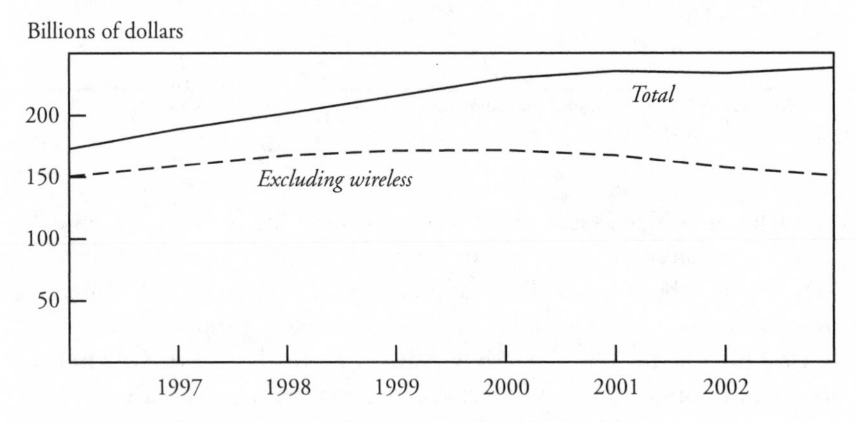

Source: FCC, *Telecommunications Industry Revenues, 2002* (March 2004).

Figure 3-5. *Output in U.S. Telecommunications, 1987–2001*

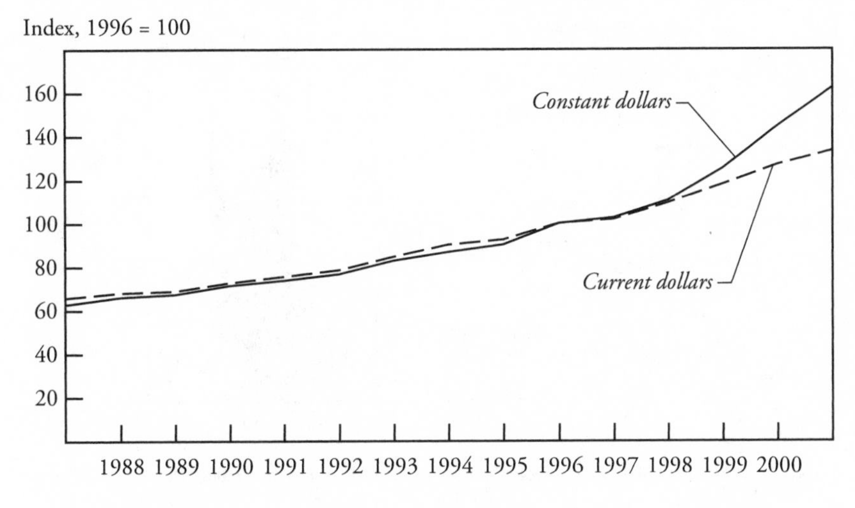

Source: BEA (2003).

from its 1987–96 level.[16] In the unregulated wireless sector, labor productivity not only grew about twice as rapidly as in the wire-based sector, but it accelerated sharply. As explained in chapter 4, the disappointing performance of wire-based firms is due largely to the entry of new local carriers, who added greatly to industry employment but not to output.

Regulatory Changes

In any newly liberalized market, many business plans will prove unsuccessful as entrepreneurs struggle to find a strategy that works. However, telecom's fast-advancing technology, soaring Internet, and lower international barriers to communications created even more opportunities for mistakes than arose during previous episodes of liberalization or deregulation, such as those in the airline or trucking industries.

For the most part, the telecom havoc of the 1990s can be traced to the following regulatory changes:

—Opening of wireless communications to competition for the first time in 1993 and deregulation of wireless rates.

—Government auctions initiated in 1995 to provide the requisite spectrum for new wireless competitors.

Figure 3-6. *Labor Productivity Growth in U.S. Telecommunications*

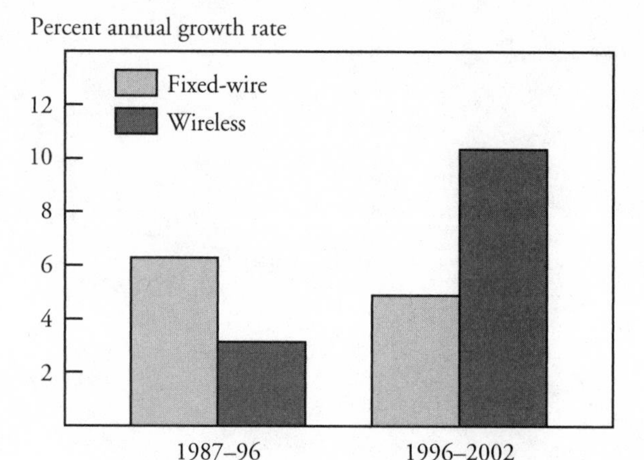

Source: Bureau of Labor Statistics, *Consumer Price Index, All Urban Households*, "Land-line intrastate toll calls," accessible at www.bls.gov/cpi/home.htm (February 2005).

—Deregulation of AT&T's interstate long-distance rates by the FCC in 1995.

—Opening of local telecommunications markets to competition by the Telecommunications Act of 1996.

—The requirement under the act that local incumbents lease network facilities to new entrants at regulated wholesale prices.

—The FCC's sweeping program launched in 1997 to reduce international "accounting rates" and to increase competition in international services.

—The replacement of other forms of competitive local entry with the "UNE Platform" as AT&T, WorldCom, and others responded to the deep discounts that states offered for leasing (essentially reselling) the incumbents' networks.[17]

All of these changes occurred just as Internet and telecommunications stocks began to soar. Equity and debt capital poured into the sector, funding many hastily constructed investment plans by firms eager to establish a presence in the market. The regulatory upheaval served to compound the problems created by this bubble.

Explaining the Slow Growth of Output and Revenues

There are at least three explanations for the revenue slump experienced by wire-based telecom carriers after 1996. First, the demand for traditional telecommunications services is price inelastic. When prices fall, total expenditures also fall unless these services are put to important new uses to generate exogenous growth, and thereby to offset the effect of price declines. Second, the revolution in wireless communications has siphoned enormous amounts of traffic from the wireline network, particularly from the overpriced long-distance traffic. Third, regulation may have reduced the incentive to invest in the facilities required to deliver risky new services that are essential to creating revenue growth.

Household Spending Growth

Approximately 60 percent of all end-user telephone expenditures are made by households. As census data show, the share of household expenditures devoted to telephone service remained remarkably constant throughout the1980s, at about 2 percent of their overall expenditures.[18] Beginning in 1993, this share rose gradually to 2.3 percent. In nominal dollars, the average household spent $877 in 2000 compared with $658 in 1993.[19] However, *all* of this increase reflected a growth in spending on wireless, not traditional wire-based, services (table 3-2). Rising expenditures on local service, reflecting principally the increase in FCC-mandated subscriber line charges, have not compensated for the decline in long-distance spending, and the result has been a sharp decline in average household expenditures on traditional telephone service since 1997.

Why have household expenditures on wire-based telephony fallen in an era of explosive growth of the Internet? The reason must be that revenues from the new uses of the telephone network have not offset the decline in revenues from rapidly falling long-distance rates.[20] Given that the residential demand for local service has an estimated price elasticity of less than –0.05, any increase in local rates would lead to higher expenditures on local services.[21] The modest increases in local rates after 1995, reflecting principally the increase in FCC-mandated subscriber line charges, did not offset the decline in long-distance spending caused by plummeting long-distance rates. The demand for long-distance services has a price elasticity of only about –0.7.[22] Therefore long-distance spending will decline when prices fall, unless the effects of higher incomes and new services are sufficient to offset the price effects.

Table 3-2. *Average Annual Household Expenditures on Telephone Service*
Dollars/year

Year	*Local carriers*	*Long-distance carriers*	*Wireless carriers*	*Total spending*	*Total nonwireless spending*
1995	358	250	82	690	608
1996	359	250	108	717	609
1997	379	305	129	813	684
1998	398	270	164	832	668
1999	402	257	205	864	659
2000	416	211	279	906	627
2001	426	176	351	953	602
2002	436	149	417	1,001	584
2003	441	122	492	1,055	563

Source: FCC, *Reference Book of Rates, Price Indices and Household Expenditures for Telephone Service* (July 2004), table 2.6.

Between 1995 and 2002, the average real price of interstate and international long-distance service fell by 51 percent, while the real price of intrastate calls dropped 17 percent.[23] Thus average long-distance prices declined by about 40 percent. Assuming a price elasticity of demand of –0.7 and no real income growth, real household spending on interstate long distance should have declined by about 14.2 percent. With real income growth of 19 percent over this period, much of this decline could have been offset by rising incomes. As table 3-2 shows, the average household's long-distance bill fell by about 40 percent between 1996 and 2002, but about 4 percent of the local-carrier bill is the increase in long-distance services due to Bell company entry into interLATA services.[24] Therefore household spending on long-distance services declined by 34 percent between 1996 and 2002, or by 42 percent in real terms. This is far more than the 14.2 percent prediction based on falling real prices. The reason for the large discrepancy, as shown in chapter 6, is that a substantial amount of long-distance spending has been diverted to wireless services.

The Dynamic Growth of Wireless

The 1996 act instituted major policy changes in the wireline telecommunications sector, but it largely ignored the wireless sector.[25] Although competition in the delivery of mobile wireless services had been limited to two carriers per local market by FCC policy since the 1970s, liberalization was thrust on this sector by the 1993 Omnibus Budget Reconciliation Act, which instructed the FCC to auction spectrum for commercial wireless uses. The first auctions were held in 1995, and construction of the new digital personal communication services (PCS) networks was just beginning when the 1996 act was passed.

The 1993 legislation virtually eliminated price regulation of mobile wireless (or "cellular") services. States may now regulate these rates only if the carriers have "market dominance," an unlikely condition in today's wireless sector, even in rural areas.[26] The auctioning of 120 megahertz of spectrum allowed four or more new entrants into wireless services in each local market to compete with the two carriers already operating there. Since 1996 mergers and consolidations have transformed the wireless industry into one with five large national carriers, including a firm—Nextel—that operates on frequencies other than those for which rights have been auctioned.[27] Two of the competitors, AT&T and Sprint, were launched by large long-distance companies.[28] One is a joint venture of two Bell companies, SBC and Bell South, and one is a joint venture of another Bell company, Verizon, and the European wireless firm Vodafone.

Although the United States launched its "second-generation" digital wireless service somewhat after Europe and Japan, the number of subscribers is now growing rapidly (see figure 3-7). By mid-2004, it had risen to almost 170 million and by 2005 is likely to exceed the number of fixed access lines.[29] The substitution of wireless for traditional wire-based telephony is clearly accelerating. Many households, particularly those with young adults, do not even have a traditional copper-wire telephone service. Others are using their wireless service rather than their home telephone for long-distance calls. The effect on traditional wire-based telephone carriers is obvious. Long-distance revenues for wire-based carriers are now declining, and the number of fixed access lines is now also falling after decades of steady growth (see figure 3-8).

The potential stumbling block for wireless operators is the technology required to provide high-speed Internet access. In Europe and other parts of the world, carriers have paid billions of dollars in auctions for spectrum designated for "third-generation" (3G) wireless services. But these services are likely to grow more slowly and be less easy to use than fixed-wire broadband services, such as cable modems and DSL, or the new wireless WiFi services. The United States has not yet auctioned additional spectrum for 3G applications, but Verizon Wireless is now offering a new high-speed service over its existing spectrum. Nevertheless, the traditional telephone

Figure 3-7. *Growth in Wireless Subscribers and Revenues, 1993–2003*

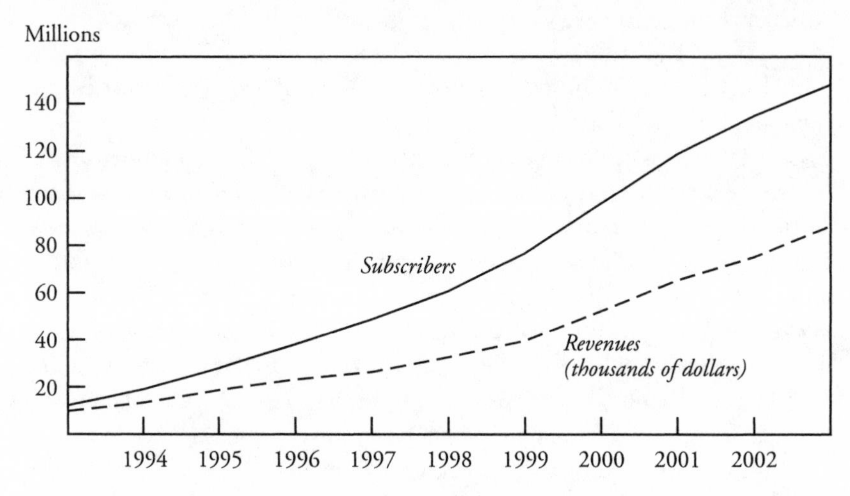

Source: Cellular Telecommunications Industry Association (CTIA), *Semiannual Survey* (www.wow-com.com/industry/stats/surveys/ [2003]).

Figure 3-8. *Growth in Switched Access and Wireless Lines, 1980–2003*

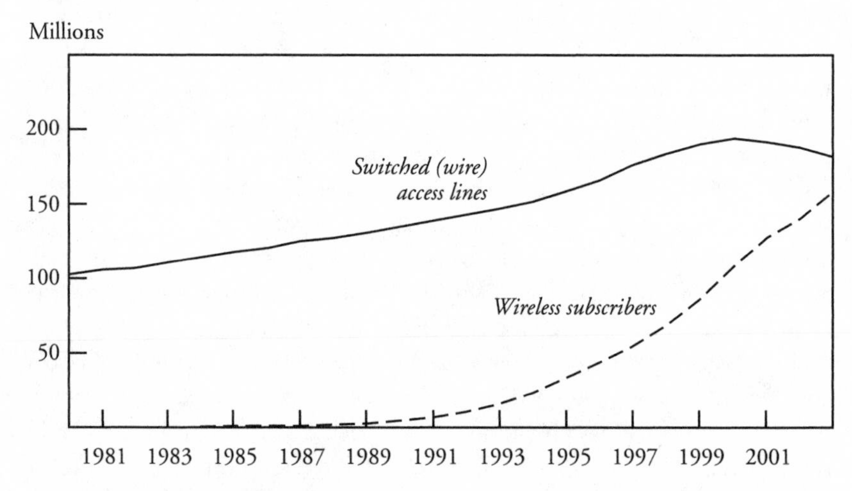

Source: FCC, *Trends in Telephone Service* (2003); CTIA, *Semiannual Wireless Industry Survey* (December 2003).

companies have an opportunity to exploit a major new revenue source if regulators will create an environment that allows them to do so.

Regulatory Uncertainty

States continue to regulate local retail rates much as they did in 1996. The regulation of wholesale rates, required by the 1996 act, is still evolving because of continuing legal challenges. The unbundling rules are still being rewritten after three previous sets of rules were overturned on appeal, and the FCC is now reconsidering its forward-looking cost rules for pricing the unbundled elements. These issues have been before the FCC for more than eight years, and the commission's 2003 ruling created further confusion and uncertainty.[30]

Similarly, the commission has been unable to provide a definitive set of rules on the provision of the new Internet broadband services, including the telephone companies' DSL services and cable television companies' cable modem services. These dual uncertainties pose particular problems for the Bell companies, whose retail revenues are shrinking because of the decline in access lines, the loss of retail lines to new competitors, and the general shift of traffic to wireless and voice over Internet protocol (VoIP) services (see chapters 4 and 5). Between 2000 and 2002, the Bell companies slashed their capital spending by 50 percent as they awaited some clarification of these issues, even though they needed to upgrade facilities to be able to deliver broadband DSL service to many of their subscribers. They are now beginning to increase network investment once again, albeit slowly.

As discussed in chapter 8, the telephone companies' DSL services compete actively with cable television systems and a variety of wireless and satellite services. Because these services have developed over facilities originally designed to carry other communications services, they could be and in some cases are subject to regulation under different provisions of the Telecommunications Act. Recently the FCC decided that cable modem service is an "interstate information service" and therefore should be under the FCC's jurisdiction, but this decision has already been overturned by a federal appeals court. The FCC is also contemplating the appropriate approach to regulating fixed-wire broadband services, including DSL, but it has been very slow in developing such a policy.[31] In addition, the sudden emergence of VoIP services has forced the FCC to begin considering whether such services are subject to the wide range of rules and charges that currently burden ordinary voice services.[32]

Through a welter of different proceedings, the commission could exercise "forbearance" from regulating any of these services under Section 706 of the act. Unfortunately, it has been examining these options for more than two years without reaching any decision on whether to regulate cable modem service or to forbear from regulating any of these "advanced" services, allowing the market to drive technology, facilities deployment, and the pricing of the services. The uncertainty this has created is surely affecting the willingness of carriers to deploy the capital required to deliver these new services and thereby expand industry revenues.

Conclusion

Virtually every developed country in the world has suffered a telecom boom-bust cycle in recent years. The exaggerated expectations created by the Internet and general IT revolution were met with stark reality in Europe, Asia, and Oceania in 2000–02 when stock market valuations plunged from the unprecedented heights of 1998–2000. With capital spending in sharp decline, most telecom equipment suppliers had difficulty surviving. The two largest North American suppliers, Lucent and Nortel, were pushed close to bankruptcy, and the market capitalization of large European firms such as Nokia, Marconi, and Alcatel dropped by more than two-thirds.

As this book goes to press, the gloom of the past four years is beginning to lift for some parts of the telecommunications sector. In the remaining chapters, I examine in more detail what went wrong and what went right in the U.S. sector since 1996 and provide some discussion of developments in other countries.

4 Local Competition under the 1996 Act

Decades of regulation have left the U.S. telecommunications sector with a distorted rate structure.[1] A substantial share of long-distance revenue has traditionally been used to defray the non-traffic-sensitive costs of the local network, through carrier access charges and excessive intrastate rates. In addition, local residential rates have been kept artificially low, particularly in less densely populated areas, while local single-line business rates are generally more than twice as high, despite the fact that the average business line is shorter than the average residential line.[2] In view of this distortion, entrants should have ample opportunity to target business and residential subscribers that use a large amount of long-distance service, particularly in the areas of greatest population density.

Even so, entrants have no road map for success because their experience with local competition is so limited. The last time the United States had competitive local telephone markets was at the beginning of the twentieth century, after the initial Bell patents expired.[3] That history would be of little use to an entrant contemplating entry into today's market, with its sophisticated modern technology.

The Difficult Economics of Local Entry

When the 1996 Telecommunications Act opened all telecommunications markets to entry, long-distance rates were still far above incremental cost, partly because of excessive carrier access charges. Given the distortion in local residential rates and the highly skewed distribution of long-distance calling, a large share of local residential customers were generating negligible net returns to local carriers.[4] But even with the profitable local customers included, incumbent local telephone carriers were not earning excess returns at this time because they were tightly regulated. At the end of 1996, financial markets valued the large Bell companies' nonwireless assets at about $1,940 per switched access line, which was approximately equal to the book value of these companies' network assets per line.[5] The value of their domestic local operations was undoubtedly even less because foreign operations, directories, and various other miscellaneous services are included in the nonwireless assets.[6] Without these assets, the value of the local exchange operations was little more than $1,600 per line at the end of 1996, or probably no more than the *reproduction* cost of network assets. Thus the tightly regulated incumbent carriers were not likely earning monopoly rents that entrants could attack after 1996.[7]

Between 1996 and 2003, the incumbents' market value grew about $230 per line, or 12 percent. During this period, the S&P 500 index of equity prices rose by about 50 percent. Therefore the Bell companies' equities increased in value by substantially less than the general stock market over this seven-year period, despite their considerable investment expenditures to upgrade their networks so as to be able to offer advanced (digital subscriber line, DSL) services (see chapter 8).

According to data from the Federal Communications Commission (FCC), the incumbent local carriers realized about $615 in annual revenue per switched access line in 1996–98.[8] The larger incumbents had cash flows equal to approximately 42 percent of these revenues, or about $258 per switched access line.[9] A new entrant might be able to exceed the incumbents' revenue yield by targeting business customers, but its capital cost per customer would likely be far above the incumbents' $1,600 investment per line in the first few years of operation because of the economies of density. An entrant that did happen to attract the lower-yielding residential customers might realize annual revenues from local service of only $300 per line or even less. At these yields, the incumbent firm would be losing money except in the most densely populated areas. Even if the entrant operated much more efficiently than the incumbent, surely a possibility, the returns would not be high enough to amortize its fixed investment if it simply replicated the incumbent's plant. An entrant would therefore have to search for a lower-cost entry strategy or provide new services to its customers so as to increase net revenues per line.

Entry Strategies

As already mentioned, the 1996 act opened local telecommunications markets to entry by barring state regulators from denying entrants the right to compete and by specifying a complex set of "interconnection" requirements between entrants and incumbent local exchange carriers (ILECs). Under the latter provisions, ILECs were required to (1) offer their services at wholesale rates to the entrants so that the entrants could "resell" them; (2) "unbundle" their networks so as to enable entrants to lease ILEC facilities, unbundled network elements (UNEs), at regulated prices that reflect costs; and (3) interconnect with entrants at any technically feasible point.[10] These provisions allowed new firms to enter local telecommunications markets without having to replicate all of the incumbents' network facilities.

Resale

Entrants have found the new entry options to be a mixed blessing. The simple resale of incumbents' services could never be a very profitable activity because under the 1996 act entrants would simply be obtaining a marketing margin equal at most to the incumbents' "avoided costs" of retailing their services. This margin between retail and wholesale prices was generally in the range of 15 to 20 percent of the retail rate because incumbents do not have to engage in large marketing programs to enroll their subscribers.[11] Hence if a residential customer was paying $360 a year for its local service, the entrant would receive a gross margin of $54 to $72 a year to pay for its marketing, billing, and other customer service costs. Entrants quickly discovered that offering residential resale service at these margins could not be a profitable stand-alone business. Unfortunately, neither is the resale of much higher-priced local business services because the entrant is locked into the incumbent's service package and cannot offer the business customer new or enhanced services.

Resale is likely to be an attractive short-term option only for an entrant whose long-range strategy is to build its own network, a strategy that takes time to implement. If the entrant markets its new services throughout a metropolitan area but its network facilities reach only a small share of the area, it can use resale to serve customers who respond to its marketing in areas that are not yet served by its own network.

In other cases, an entrant may be offering a bundle of services over its network that does not include local telephony. For example, RCN has built hybrid fiber-coaxial cable networks to deliver broadband Internet access and video services. If it does not wish to build its own circuit-switched telephone facilities into such a network, preferring to wait for an improvement in the technology for delivery of voice over Internet protocol (VoIP) services, it may decide to offer resold local telephony service to fill out its "bundle" of services.

Network Unbundling

A very large share of the incumbent local telephone network's $1,600 to $2,000 in investment per line is sunk in the distribution plant deployed to reach the final subscriber. Replicating this cost for all but the highest-revenue customers in the more densely populated areas is a daunting task unless the entrant can offer an array of new services. For this reason, the 1996 act included a requirement for unbundling incumbent facilities and providing them at wholesale "cost-based" rates to entrants. These rates are established through state-by-state arbitration and regulation and have been the subject of endless controversy since the act was passed. Despite the FCC's prescription that these rates be based on forward-looking, total-element, long-run incremental cost (TELRIC), the actual rates vary enormously across the states. For example, the rate for leasing the subscriber line in the areas of greatest population density in Illinois was just $2.59 to $7.07 a month in early 2002, but it was more than $15 in such areas in eleven other states.[12]

TELRIC prices are supposed to be based on the cost of building new facilities and amortizing them over their full economic life, but entrants are allowed to lease the incumbents' facilities at these rates on a month-to-month basis. Therefore for facilities that are irretrievably "sunk," entrants are provided with a free option because they do not have to incur the risk that the incumbent faces from technological obsolescence or other forces that could strand such an investment.[13]

The requirement for network unbundling can be defended on grounds similar to those just advanced for resale. Building network facilities to connect all of an entrant's customers can be extremely expensive and time-consuming. An entrant could begin by investing in local switching and interoffice transport capacity but use the incumbents' local loops to connect its customers to its own switches until it has enough of them to justify building its own distribution lines in densely populated areas. Doing so may never be warranted in the less densely populated areas if the entrant is using a technology based on terrestrial wires or cables. In these latter areas, it may almost always be necessary to rely on unbundled loops or resale. Hence these may be "essential facilities" to which the incumbent is required to provide entrants access.[14]

The use of unbundled network elements grew substantially once regulators began insisting that the entire complement of incumbent facilities be provided to entrants at these TELRIC rates, thereby essentially providing the entrant with the opportunity to resell the incumbent's services at a much larger wholesale discount.[15] If the entrant used this entire UNE "platform," however, it sacrificed the flexibility to design its own network to offer new services. In other words, it simply obtained a resale option at a wider discount.

Facilities-Based Entry

An entrant into local telecommunications may choose to build its own network facilities, thereby allowing it the flexibility to develop new technologies or offer new services. These facilities may be in the form of the fiber-optic/copper plant deployed by the incumbents, fiber-optic/coaxial cable deployed by cable television companies, fixed wireless facilities, cellular systems, or even satellite systems. While such entry is initially more expensive, it enables the entrant to differentiate its services from those of the incumbent. Equally important, it frees the entrant from the need to depend on its principal rival for maintaining and modernizing its network.

Growth of the New Entrants

Before 1996 there was little local mass-market competition in the U.S. telecommunications sector because state regulators generally discouraged local competition. A few competitive access providers (CAPs) had begun to build fiber rings in central business districts principally to interconnect large customers with long-distance carriers. State regulators could not block these entrants because they were providing a local service defined as "interstate access." However, the 1996 act opened all telecom markets to entry, including the local market for small business and residential subscribers.

Entry began slowly in 1997–98, as total lines of competitive local exchange carriers (CLECs) increased from about 1.5 million CAP lines to

Figure 4-1. *U.S. Local Competitors' Share of Access Lines, 1997–2004*

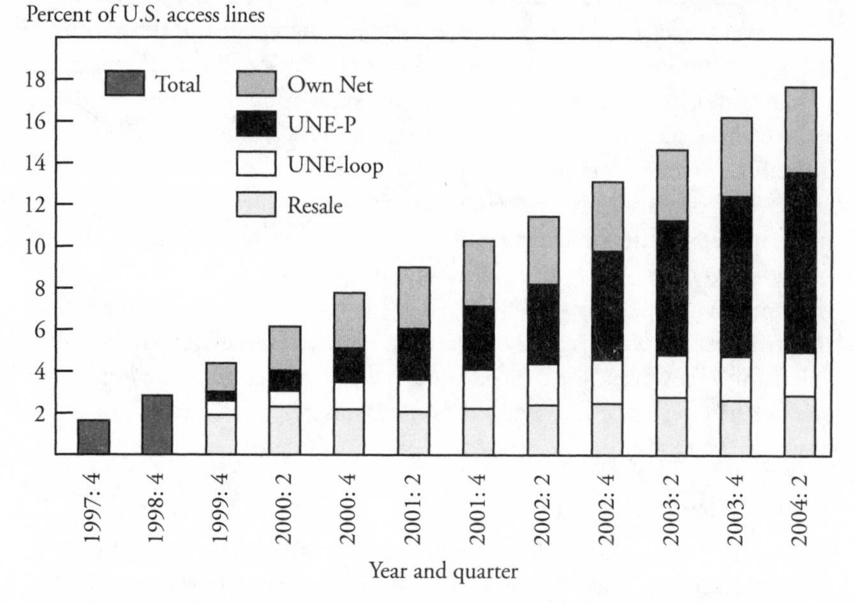

Source: FCC, *Local Telephone Competition: Status as of December 31, 2003* (June 2004).

2.4 million total entrant lines. Thereafter the pace accelerated, and by the middle of 2004 entrants had 32 million lines, or about 17.8 percent of all switched access lines (figure 4-1).

Network Strategy

More than two-thirds of the growth in CLEC lines since June 2000 is attributable to resale, including traditional, total service resale and the lines leased as UNE-P (figure 4-1). As might be expected, the share of CLEC lines represented by traditional resale has not grown appreciably since the end of 1999. The 15 to 20 percent gross margins available to the entrants from this form of entry are not very attractive. However, the use of the UNE-P has soared, in large part because entrants have been able to gain access to the entire platform at low TELRIC rates or even less without having to make any capital expenditures.[16] This use of the entire incumbent's network is, in essence, resale at discounts of 50 percent or more, providing potentially very attractive margins for arbitrageur-entrants (see table 4-1).

Table 4-1. *Arbitrage Margins Available in Bell Company Territories, 2002*

Company	*Average wholesale UNE-P rate (dollars/month)*	*Average retail revenues/line (dollars/month)*
BellSouth	23.10	53.69
SBC	16.55	51.23
Verizon	19.40	42.49
Qwest	22.94	51.10

Source: Anna Maria Kovacs, Kristin Burns, and Gregory S. Vitale, *The Status of 271 and UNE Platform in the Regional Bells' Territories* (Commerce Capital Markets, November 8, 2002).

As a result, by the end of 2003 nearly two-thirds of all entrants' lines reflected little more than the resale of the incumbents' services.[17] Before 2001 the share of CLEC lines served by their own facilities had been growing, but it had receded to just 23 percent by the end of 2003. The rest of the 30 million entrants' lines were leased or resold from the incumbents.

Even more significant, reliance on the unbundled platform has been associated with a lack of growth in the new noncable entrants' use of their own lines. The slight increase in the net share of all entrants' own lines since June 30, 2000 (see figure 4-1), reflects the steady expansion of telephone subscribers to cable television systems, from 1.2 million in December 2000 to 3.3 million in June 2004. By contrast, the facilities-based lines of noncable company entrants remained constant at about 4 million lines.[18] Thus it appears that noncable entrants stopped investing in their own facilities, perhaps because of the adverse outcomes for those who did so before the 2000 collapse in CLEC stock values, but also because the environment created by regulators provided passive resellers of incumbent services more attractive returns than did investing in their own switches or even their own complete networks.

The growth of entrants using their own switches and leasing incumbent loops has also slowed with the expansion of the resale of incumbents' services through the UNE-P. This trend would not likely have reversed itself under any conditions because there is no evidence that the entrants who use the entire UNE-P—such as MCI, AT&T, Talk America, or Z-Tel—were doing so to obtain a "toehold" before launching facilities-based entry. Now that the Court of Appeals has virtually closed the door on the UNE-P, however, these companies will be forced to invest in at least some of their own facilities or slowly withdraw from offering local services. Most will choose the latter course.

The Effect of UNE Rates on Entrants' Choice of Network

Given the enormous variation in wholesale UNE rates across states, it is possible to test for the effect of these rates on the CLECs' choice of network strategy in the first five years of the new act's implementation. If the UNE rate is lower than construction cost per line, CLECs would obviously want to lease lines rather than build them. On the other hand, if the UNE rate exceeds construction costs, CLECs would be motivated to build their own facilities, all other things being equal.

To determine the effect of UNE rates on the use of unbundled loops, I and two colleagues undertook an econometric analysis of the choice of network design in 2000–01 and 2002.[19] First, we estimated the direct elasticity of substitution between UNEs and the CLECs' own loops from a cross section of state data:

$$log(F_i) - log(U)_i = a + b^*log(UPrice_i) + d^*log(FCost_i) + u_i, \qquad (1)$$

where i is a geographic (state) index, F = CLEC-facilities-based lines, U = UNE lines, *Uprice* = statewide average UNE loop rate, *Fcost* = statewide average of FCC ARMIS data on embedded costs per line or FCC estimates from the Hybrid Cost Proxy Model (HCPM), and u = random disturbance term. We expected that the absolute value of the estimate of b would not be significantly different from d. If the estimates passed this empirical test, we could then estimate equation (2), to obtain an estimate of the direct elasticity of substitution, b:

$$log(F_i) - log(U_i) = a + b^*[log(UPrice_i) - log(FCost_i)] + e_i\,. \qquad (2)$$

For the monthly UNE price per line, we used the average UNE loop rate as determined in proceedings supervised by each state's Public Utilities Commission (PUC) and published by the National Regulatory Research Institute (NRRI).[20] To estimate the cost of building loops, we used two alternatives: the embedded cost per loop from the FCC's ARMIS data and the FCC's HCPM model. We calculated the number of CLEC facilities-based lines and the number of UNE lines from two sources as well: the E911 database and the FCC's *Local Competition* reports. Thus our model focuses on a CLEC's incentive to lease UNE loops in relation to investing in its own facilities-based network, a decision that most CLECs face.[21]

In a least-squares regression estimate of equation (1), using the available data on CLEC lines, we found the coefficient of the *FCost* variable to be negative as expected, and the absolute values of these two regression coefficients in equation (1) were not significantly different.[22] The resulting estimates of the direct elasticity of substitution in equation (2) are generally statistically significant and between 0.4 and 1.4, depending on the year and the source of the CLEC lines data. These estimates suggest that wholesale UNE rates have a substantial effect on the incentive for entrants to invest in their own facilities.[23]

The Determinants of the Level of CLEC Entry

Given the short history of local competition, there is not a wealth of evidence on what influences local entry. Even the few empirical studies published on the subject shed little light on the determinants of *successful* entry because most of the local entrants examined were failures. In general, entry appears to correlate with market size and density, surely a reasonable result.[24] Entrants have tended to target business customers more than residential customers for obvious reasons: regulated business rates are much higher than residential rates. Even if they are not using their own network facilities, entrants will be attracted to the larger, urban markets because of the economies of density in marketing their services. Entry is also correlated with population or employment in the local market.[25] Interestingly, many entrants avoid the largest markets because of the difficulty in competing with the first movers in this arena, the CAPs.[26]

Entrants fall into several types: some are national, others local or regional. Many of the latter are privately owned firms about which little public financial information is available. Some serve business customers exclusively, others both businesses and residences. Some target a particular group of small to medium businesses. Because entry conditions have been greatly eased by providing entrants access to incumbents' facilities, a major problem for the new competitors is their proliferation in a given market. This is particularly true of the larger, national CLECs in the major metropolitan areas. For this reason, the most successful entrants may be those who target a particular class of customers in medium-sized cities that are still unexplored by the larger national entrants.[27]

So far scholarly analysis of the effect of state regulation on the extent of local entry has provided mixed results. Some studies have found that rate-of-return regulation reduces UNE-based entry, because regulated UNE rates tend to be higher in such states, but that price-cap regulation tends to produce more entry in general.[28] Using *number of entrants* rather than extent of penetration to measure entry, one study has shown that local access and transport areas (LATAs) in states with price-cap regulation attract fewer entrants than those without such incentive-based regulation.[29] This model does not include wholesale or retail prices and does not account for the effects of cable telephony or wireless competition. Still others report that UNE rates, as well as the resale discount, have little effect on total CLEC lines when the effect of Bell company entry into long distance is taken into account.[30] However, they do find a mild positive effect on entry from the regulated retail residential rate.[31]

At this stage, as mentioned earlier, no one really knows what drives *successful* entry. In the first eight years, entrants typically offered virtually the same services as the incumbents did, particularly to residential customers. They also went after small to medium-sized businesses, but their instability made them dubious candidates for providing such an essential service to this market. The next section documents the entrants' struggle.

How Have the Entrants Fared?

Given the limited experience with rivalry in local telephone markets and the time it would take to develop entry strategies, it is hardly surprising that entry has not developed at a more rapid pace. In 1996 capital spending by the local entrants, excluding the CAPS, was only $1.2 billion, but it increased rapidly thereafter, more than doubling each year through 1999.[32] In 2000 entrants reportedly spent $12 billion on capital facilities.[33] Unfortunately, despite these expenditures, few of them were able to attract subscribers rapidly, so by 2000 they were beginning to encounter severe financial difficulties.

Capital Spending and Equity Values

By the end of 1999, entrants, excluding the CAPs, had spent at least $22 billion on capital facilities, but they had attracted only 8.2 million subscribers, 3.5 million of which derived from the low-margin resale of incumbent services.[34] Even ignoring the CAPs' lines altogether, the new entrants' cumulative 1996–99 capital spending amounted to $4,700 per

Figure 4-2. *Market Value per Line for RBOCs and CLECs, December 31, 1999*

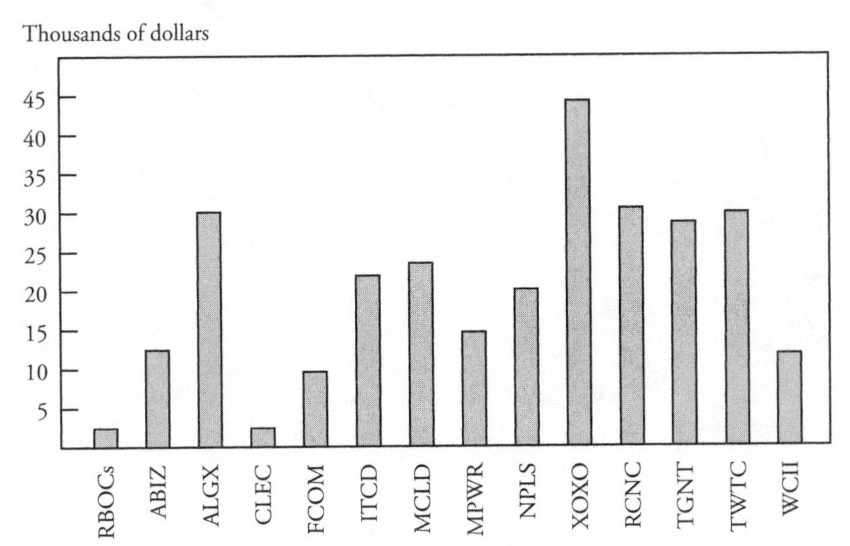

Source: Author's calculations and www.quote.yahoo.com.

UNE or facilities-based line, or $2,700 per total line, including the lines offered through simple resale.

At end of 1999, the publicly traded CLECs were valued at an astounding $116 billion, or about $20,000 per switched access line (excluding the CAPs).[35] By contrast, as figure 4-2 shows, the weighted average of Verizon, Bell South, and SBC market values per switched access line (excluding their wireless operations) was less than $2,500, or about one-twelfth the value of the leading CLECs at the time, namely, Allegiance, Time Warner Telecom, XO, McLeod, Teligent, and RCN. Obviously, the financial markets had a very optimistic view of the future prospects of these new local entrants in late 1999. This view would change with a vengeance by the middle of 2000.

When the market value of telecom firms dropped sharply in 2000–02, CLEC equities took the steepest dive, driving the value of the listed CLECs from $116 billion to just $10 billion (see figure 4-3). The total value of the surviving CLECs was about $1,300 per line at the end of 2002.[36] By the end of 2003, the surviving CLECs were valued at only about one-quarter

Figure 4-3. *Local Telecom Company Stock Prices, 1996–2004*

Index, February 1996 = 100

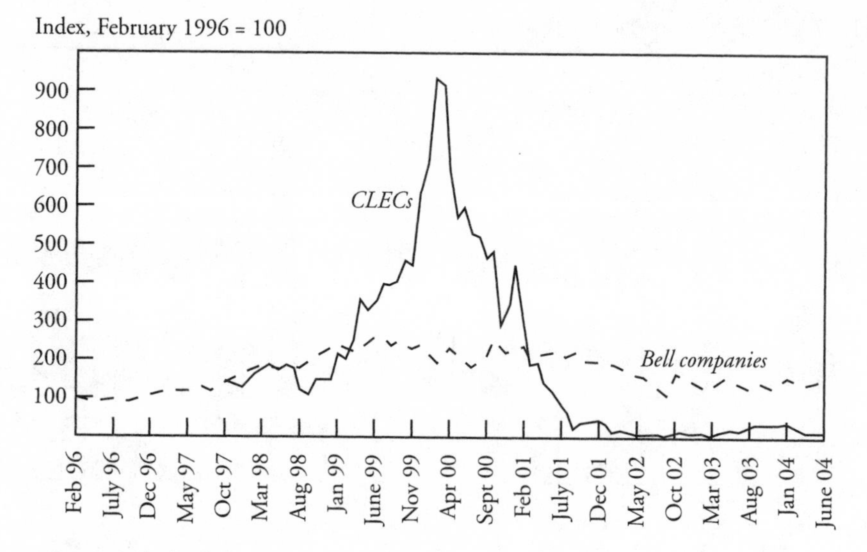

Source: Author's calculations and www.quote.yahoo.com.

of all listed CLECs' cumulative capital expenditures since 1996.[37] The equity markets reacted harshly when it became clear that these companies are not likely to grow very rapidly in the foreseeable future.

As discussed in the next section, the long-distance carriers (IXCs) also plummeted in value, whereas cable and regional Bell operating companies (RBOCs) experienced a more modest decline, generally outperforming the overall equity market. The telecom "meltdown" (see figure 3-2) therefore affected primarily the long-distance carriers, the wireless companies, the wholesale fiber suppliers, and the CLECs (the travails of the first two are examined in chapters 6 and 7).

Subscriber Revenues

One of the reasons why the CLEC market valuations shown in figure 4-2 did not persist is that these new companies were unable to generate revenue growth from their modest subscriber base. As the CLECs grew, their end-user revenues per subscriber fell steadily (see figure 4-4). Apparently these entrants were unable to attract many large business and institutional customers or to "cherry-pick" residential customers with large monthly bills. As a result, the CLECs' average annual end-user rev-

Figure 4-4. *ILEC and CLEC End-User Revenues per Line, 1997–2003*

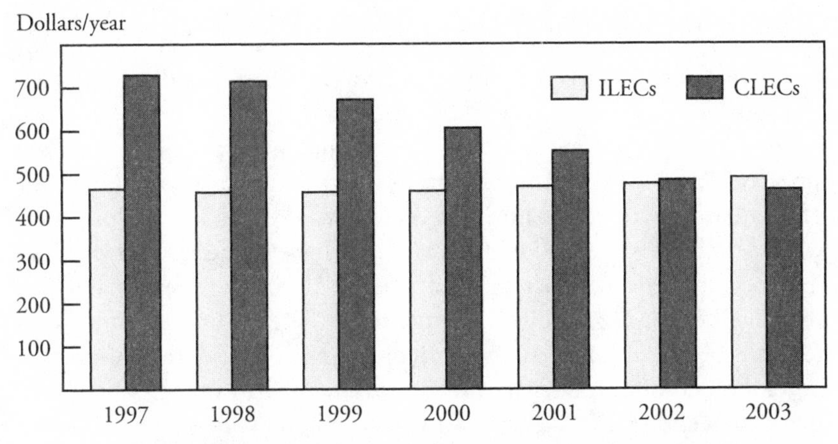

Source: FCC, *Telecommunications Industry Revenues, 2002* (March 2004); FCC, *Local Telephone Competition: Status as of December 31, 2003* (June 2004). Revenues per line are equal to the full year's revenue divided by the average of December 31 end-user lines data for the current and preceding years.

enue per line fell by more than $150 between 1997 and 2001, a period in which ILEC revenues per line increased slightly. These trends could not have been very heartening to entrants, or to their investors, who must have thought they could exploit the distorted regulated rate structure to gain a toehold. Moreover, the Bell companies' huge advantage in broadband is likely to narrow the differential substantially.

Assume that in 1999 investors could have foreseen that the entrants' *total* revenues per line, including wholesale revenues, would decline to the level they achieved in 2002, or $689 a year.[38] Also, assume optimistically that they could achieve cash-flow margins of 42 percent (equal to margins of the incumbents). In that case, the CLECs' cash flow would have averaged about $289 per line in 2002. Even then, to justify the market capitalization of $20,000 per line achieved at the end of 1999, these carriers would have had to increase their subscribers at a rate of 24 percent a year ad infinitum, assuming a before-tax cost of capital of 25 percent.[39] But a 24 percent annual growth in lines would find the entrants owning all of the country's lines in just thirteen years, not a likely outcome.

Alternatively, one could assume no subscriber growth and an annual cash flow percentage equal to that of the incumbents, in which case these CLECs would have had to have more than 30 percent of the country's access lines at the end of 1999 to justify their market capitalization at the *RBOCs' cost of capital*! But at that time, the entrants had only 4 percent of the country's access lines and a much higher cost of capital. Clearly, the equity markets had placed a very optimistic set of bets on these companies in 1999.

Why have CLEC end-user revenues per line generally been declining since 1997? Given the distorted local rate structure, one would have expected entrants to target business customers and residential customers who spent heavily on telecommunications. However, nearly two-thirds of the entrants' lines connect to residences and small businesses, and this share has been rising steadily.[40] Furthermore, these are not residential customers whose telecom spending is atypically high.[41] Also, the entrants do not appear to offer residences much lower rates than the ILEC rates for their connections. The average CLEC local bill in the first two quarters of 2001 was $35 a month for CLEC customers and $32 a month for ILEC customers.[42]

A linear regression of the monthly local bill on the various calling features and the number of lines finds that the prices charged by CLECs for calling features, such as call waiting or caller ID, are strikingly similar to those charged by the ILECs in the first two quarters of 2001.[43] After netting out these features and the number of lines, the average charge for local residential services was slightly higher for entrants ($25) than for incumbents ($23). The average number of features chosen by CLEC customers (2.25) was slightly lower than the number chosen by incumbent-carrier customers (2.40). In short, the entrants did not attract households with a greater propensity to spend on telecom services, nor did they appear to offer residential subscribers measurably lower rates. In light of these facts, one might reasonably ask why these subscribers bothered to shift to CLECs at all. Of course, the rates offered *business* subscribers may have been substantially below those offered by the incumbents.[44]

Given the limited penetration of the new entrants into the residential market, it is difficult to estimate a model of consumer choice of local carrier through an econometric model. In addition, it is difficult to ascertain the availability of CLEC services in various parts of the country.[45] How, then, can one determine if a household that subscribes to an incumbent carrier's services makes this choice because there is no alternative or because it is unaware of any alternative? To explore this question, I use a logit model of household choice between a CLEC and an ILEC as its local carrier for the first and second quarters of 2001. I focus on urban households since entry in urban areas was advanced enough by 2001 to provide a very large share of households in these areas with at least one competitive choice.[46]

The model's independent variables are size of the metropolitan area, income, age of the head of household, number of persons in the household, race or ethnicity, and two dummy variables indicating whether the household subscribes to AT&T or MCI for its long-distance service. The latter two variables are intended to capture the fact that AT&T and MCI are also new entrants into local exchange service and may therefore have an advantage in inducing their customers to switch from the incumbent local carrier.

Because of the large number of local entrants, I was unable to identify a single price variable for the entrant's local service for each household in the sample. A given household may have a number of entrants to choose from, and their prices may vary. Instead of using any one entrant's price, I use the UNE-P wholesale rate that any entrant faces in the state in which the household resides. For the incumbent rate, I use the average ILEC rate in the state, stripped of the estimated prices of second lines and calling features. Hence my price variable is the ratio of the UNE-P rate to the average ILEC rate for the state.

For the first quarter of 2001, my sample includes 658 households, 52 of which used a new entrant for their local service. The chi-square goodness of fit score for the entire logit regression is 50.3, which is significant at the 2 percent level of confidence. The relative price coefficient is –0.379, which has the expected sign but is not statistically significant. For the second quarter of 2001, with 49 households of a total of 685 subscribing to an entrant's service, the goodness of fit improves to a chi-square of 57.6, which is significant at the 0.4 percent level of confidence. The estimated coefficient for the UNE-P/ILEC price variable in this quarter is –2.094, which is significant at the 0.5 percent confidence level. Thus in the second quarter of 2001, at least, households appear to have been responsive to relative prices of local service. At the point of means, the elasticity of CLEC choice with respect to relative prices is close to –2.0.

Why Have Local Entrants Struggled?

The market's lofty expectations for CLECs have clearly not been realized. What has happened since 1999 to suggest otherwise? As explained earlier, the simplest answer is that CLECs could not and did not wrest sufficient revenues from the incumbents to satisfy these expectations. By the end of

1999, it was already clear that their revenues per line were declining steadily (see figure 4-4). At that time, 41 percent of CLEC lines were residential and small business lines; by the end of 2003, this share had risen to 63 percent. Between December 1999 and December 2003, the incumbent companies' line counts among medium and large businesses and institutions fell by 8.4 million, while the CLECs attracted 6 million net additions from these large users.[47] But there is little evidence that the CLECs were attracting the high-revenue customers from this group.

In earlier research, I concluded that the entrants' network strategies were an important determinant of their ability to grow and to attract revenues.[48] Specifically, CLECs that built their own networks were more likely to convert capital spending into revenues than were those that relied heavily on resale or UNEs. Since that time, however, several of the larger facilities-based carriers, such as XO and RCN, have encountered severe difficulties and have been reorganized under the protection of bankruptcy. Indeed, all of the new carriers are now struggling or have failed altogether.

By the end of 2002, the market values of new entrants, regardless of network strategy, were substantially less than the cumulative 1996–2002 capital expenditures by all new entrants in each category (figure 4-5). By this measure, the carriers building their own facilities had slightly outperformed those relying on UNEs, but the difference was very small. Since reported capital expenditures presumably do not include start-up losses, all three groups of entrants obviously performed badly by this measure.

Surprisingly, reliance on unbundled elements or resale does not generate more revenues per dollar of capital investment than building one's own facilities. The carriers that survived into 2002 by relying heavily on UNEs generated $0.67 in 2001 revenues for each dollar of cumulative capital spending between 1996 and 2001.[49] Those relying on resale did somewhat worse, obtaining $0.47 in 2001 revenues per dollar of cumulative investment.[50] But those investing in their own facilities generated $0.65 in 2001 revenues per dollar of cumulative investment, presumably because by having their own facilities they could generate more revenue per customer than resellers to offset their greater investment.[51]

By a second measure, the change in market capitalization, all CLECs have suffered. The facilities-based carriers' market values declined by more than 86 percent between December 31, 1999, and December 31, 2003. During this same period, resellers lost 95 percent of their market value, and UNE-based carriers lost 91 percent of theirs. The facilities-based carriers and the UNE-based carriers each had total market capitalization of $36 bil-

Figure 4-5. *CLECs' Market Value, December 31, 2002, versus Cumulative Capital Expenditures, 1996–2002*

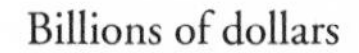

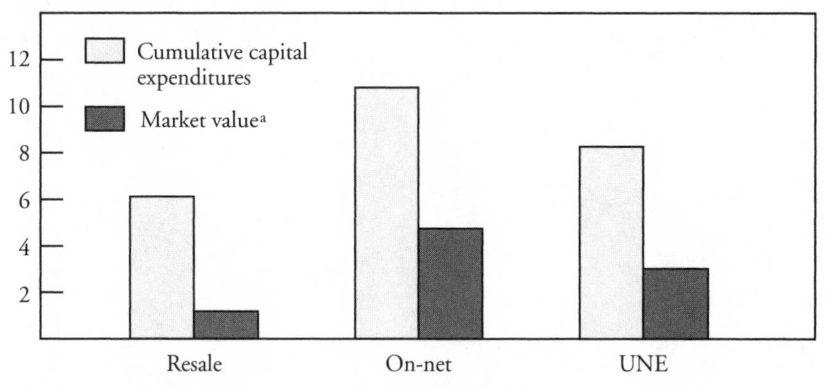

Source: Company reports; and www.quote.yahoo.com.
a. The market values include the book value of debt.

lion on December 31, 1999. By the end of 2003, these values had declined to just over $2 billion. Over the same period, the resellers' market cap fell from $20.3 billion to $0.4 billion. In short, there was very little equity left in the competitive local exchange sector by the end of 2003.

A closer look at the individual entrants provides little evidence of a winning strategy among any of them. Of the major publicly traded CLECs that entered in 1996–2002 (excluding the long-distance carriers), few have survived, and of the survivors, many have been forced into bankruptcy (table 4-2). Twenty-five of these survivors, excluding WorldCom and AT&T, realized revenues of $9.3 billion in 2002, but these sums include a variety of nontelephone revenues. At the end of 2002, cable television companies had 3 million telephone subscribers, or 0.8 million more than at the end of 2001.[52] Assuming that these cable companies realized $500 per line in telephone revenues, they accounted for about $1.3 billion in CLEC revenues in 2002. Given that the competitive local carriers realized $18.5 billion in total revenues in 2002, as reported by the FCC, the CLECs not listed in table 4-2, namely, the long-distance carriers and various privately traded companies, accounted for about $7.9 billion in revenues in 2002.

In addition to the publicly traded companies, about forty small, private firms were offering telecommunications service at the end of 2002. These companies accounted for about one-eighth of entrants' telephone revenues

Table 4-2. *CLECs: Entrants and Survivors, 2004*

Company	*Network strategy*	*Survivor?*	*Revenue (millions of dollars, 2002)*
Adelphia Business Solutions	Resale/on-net	Yes (operating in bankruptcy)	. . .
Allegiance Telecom	UNE	No	771
Allied Riser	On-net	No (acquired by Cogent)	. . .
Advanced Radio		No	. . .
US LEC Corp	UNE	Yes	247
Choice One	UNE	Yes	291
Cogent	On-net	Yes	52
Concentric Network	On-net/UNE	No	. . .
AT&T	UNE-P	Yes	Local revenues unknown
ATX (formerly CoreComm Ltd)	Resale	Yes	299
Convergent Communications	Resale	No	. . .
Covad Communications Group	UNE	Yes (reorganized in bankruptcy)	384
CTC Communications Corp.	Resale	No (acquired by Columbia Ventures)	330
CapRock	Resale	No (acquired by Mcleod)	. . .
Cypress Communications	UNE plus on-net	No (acquired by US RealTel)	. . .
DSLNet	UNE	Yes	46
Elec Communications	UNE-P	Yes	14
Electric Lightwave	UNE/on-net	No (acquired by Citizens)	185
e.spire Communications	UNE	No (acquired by Xspedius in bankruptcy)	. . .
Ixnet	UNE	No	. . .
Focal Communications Corp.	UNE	No	329
GCI	UNE/on-net	Yes	32
GST Telecom	UNE plus on-net	No (acquitted by Time Warner in bankruptcy)	. . .

ICG Telecommunications	UNE/on-net	Yes	420
Intermedia Communications	UNE	No (acquired by WorldCom)	. . .
ITC DeltaCom	UNE	Yes	415
Level 3 Communications	Owns 1/3 of RCN	Yes	1,101[a]
McLeod USA Inc.	Resale plus UNE	Yes, reorganized in bankruptcy	992
Mpower Holding Corp.	UNE	Yes	146
Network Access Solutions	UNE	No (acquired by DSLNET)	50
Network Plus CP	UNE	No (assets sold in bankruptcy)	. . .
NorthPoint Communications	UNE	No (assets sold in bankruptcy)	. . .
Net 2000 Communications	UNE plus resale	No	. . .
XO Communications (Nextlink)	On-net/ UNE	Yes (reorganized in bankruptcy)	1,260
Pac-West Telecomm	UNE	Yes	164
Pointe Communications	UNE	No	. . .
RCN Corp.	On-net/resale	Yes (in bankruptcy)	457
RSL		No	. . .
Rhythms NetConnections	UNE	No	. . .
Talk America	UNE-P	Yes	318
Teligent	On-net	Yes (reorganized in bankruptcy)	. . .
Telocity	UNE	No	. . .
Time Warner Telecom	On-net	Yes	696
USOL Holdings	On-net/UNE	Yes	16
World Access		No	. . .
Winstar Communications	On-net/UNE-P	No (sold in bankruptcy to IDT)	. . .
WorldCom	UNE-P	Yes (reorganized in bankruptcy; now called MCI)	Local revenues unknown
Z-Tel Technologies	UNE-P	Yes	235
Total for public companies		(25 public companies)	9,250

Source: Financial reports of companies; also www.quote.yahoo.com and www.alts.org.

a. Communications revenue only.

in 2002; virtually all of this was from the use of UNE loops or the UNE platform (see table 4-3).[53] These private companies appear to have a narrower geographical focus, generally using unbundled loops to serve small to medium-sized businesses. Without public financial data on their performance, it is impossible to know if they are more successful than their publicly traded counterparts listed in table 4-2. It is important to stress, however, that they apparently accounted for fewer than 2 percent of U.S. switched access lines in 2002.

Unless entrants can attract high-revenue customers or offer other customers new services, they will have difficulty competing with the incumbent telephone companies. The recipe for success may be facilities-based entry that allows the entrant to offer an array of services—such as video, long-distance telephony, local telephony, and high-speed Internet access—over a single, integrated platform. Such a strategy may provide the entrant with an advantage over its regulated incumbent rivals, many of whom were unable to offer long-distance or other services that cross LATA boundaries or have an old copper plant that cannot deliver high-speed services to a large share of their subscribers, and all of whom have the provider of last resort responsibility at low regulated local rates. This is precisely why cable companies are expanding and why the noncable CLECs that do not have their own facilities are struggling or even collapsing.

In addition, there may be limited room for smaller carriers providing niche services or better service to businesses than the large, regulated incumbents can offer. This may explain why there are still a large number of smaller, private CLECs in operation today. But the roster of the publicly traded companies, each of which surely is cognizant of the strategies of some of these smaller, private companies, provides no obvious recipe for success.[54]

Local Entry by Long-Distance Carriers (AT&T and MCI-WorldCom)

The two largest long-distance carriers, AT&T and MCI (formerly WorldCom), are major participants in the delivery of local telecommunications services. By June 2003, each was offering approximately 3 million subscriber lines through the UNE platform, and by March 2004 AT&T alone had 4.4 million access lines.[55] These carriers have been highly aggressive in lobbying for low UNE-P rates, particularly since the Bell companies have gained access to the interLATA long-distance market.

Table 4-3. *Private U.S. CLECs, End of 2002*

Company	*Network strategy*	*Current status*
KMC Telecom	Facilities and UNE	Solvent
Birch Telecom	UNE	Emerged from bankruptcy; operating
Pac Tec Communications	Unknown	Unknown
Ionex Telecom	Unknown	Unknown
BTI Telecom	Facilities; UNE	Solvent
Globalcom	UNE	Solvent
Broadview Networks	Unknown	Solvent
New South Communications	UNE	Solvent
Integra Telecom	UNE	Solvent
Eschelon Telecom	UNE	Solvent
NuVox Communications	UNE	Solvent
Eagle Comm	Unknown	Solvent
U.S. Telepacific	UNE	Solvent
Florida Digital Networks	UNE	Solvent
Advanced Telecom Group	UNE	Solvent
Network Telephone	UNE	Solvent
NTS Communications	UNE	Solvent
Knology	Facilities	Bankrupt; operating
RNK Telecom	UNE	Solvent
Buckeye Telesystem	Facilities	Solvent
Xspedius	UNE	Solvent
Global NAPs	UNE	Solvent
Lightship Telecom	UNE	Solvent
Conversent	UNE	Solvent
Grande Communications Networks	Facilities	Solvent
Sigecom	Facilities	Solvent
Everest Connections	Facilities	Solvent
RIO Communications	UNE	Solvent
Approximately 10 others	Various	Solvent

Source: Various public sources and press releases.

In December 1999, Verizon was granted permission to offer long-distance services in New York. SBC gained entry in Texas in 2000. Since that time the Bell companies have been granted approval to offer in-region interLATA services in the rest of the lower forty-eight states and the District of Columbia and now can compete for long-distance customers in all of these states.[56] Combined with the aggressive wireless competition that has developed, this entry has placed enormous pressure on the large long-distance companies that are also CLECs to respond by offering a bundle of local and long-distance services in these states.[57] As a result, following Bell entry the CLECs' share of lines in the state usually rises much more rapidly than the national average.

When the national growth in CLEC lines is compared with that in three of the first states to allow Bell company entry into long distance (Texas, New York, and Massachusetts), each of these states registers a sharp rise in CLEC lines following the Bell company's long-distance entry (figure 4-6). In Texas, virtually all of the entry has been accomplished through UNE-Ps, while in New York approximately 80 percent of the CLEC lines are UNE-Ps. More than half of all UNE-P lines through December 2002 can be found in these two states, which led the way in granting their incumbent carriers the right to compete in long distance.[58] Only Massachusetts shows little evidence of reliance on the UNE platform, perhaps because of its high wholesale rate.

Recent research shows that the number of entrants using UNE loops in a state is directly related to Section 271 approval for the RBOC in the state.[59] In one case the number of such entrants rose from five to eight carriers in the year in which the Bell company gained approval for entry into interLATA long-distance service in the state. However, the number of entrants using UNE loops is not a good indicator of the degree of sustainable competition. Many of the public firms using UNE loops have failed or are failing, and the private companies are very small in general. These findings may reflect the fact that states ultimately use the *appearance* of competition, in the form of UNE-based entry, to justify Section 271 approval, not that such approval creates sustainable entry. Indeed, as I have just shown, the number of lines accounted for by entrants generally rises *following* Bell company entry into long distance as MCI (WorldCom) and AT&T accelerate their use of the UNE platform to try to stave off Bell company competition for long-distance customers.

One may quibble over whether entry by the long-distance carriers through the use of the ILEC platform is competition in any real sense. Recall that I was unable to find any effect of CLEC entry on local residential rates through mid-2001.[60] The extent of this type of competition may be limited because the entrants have little new to offer through this extension of resale. But even if the CLECs could continue to enroll customers through the use of the UNE platform, they would not likely prosper by

Figure 4-6. *Effect of Bell Company entry on CLEC Competition, 1999–2003*

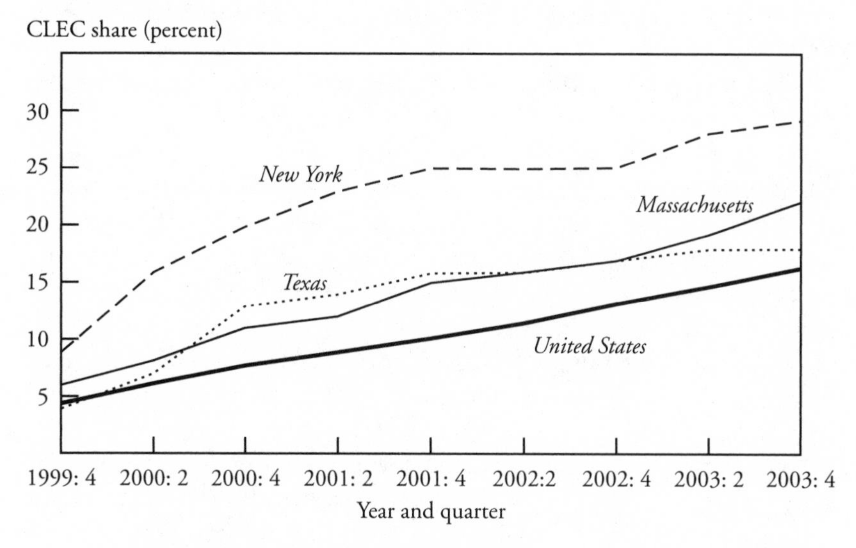

Source: FCC, *Local Telephone Competition: Status as of December 31, 2003* (June 2004), Table 7.

doing so. Anyone can resell the incumbents' facilities in this way; indeed, arbitrageurs such as Birch, Z-Tel, Talk America, and IDT (through its Winstar division) were already doing so before the Court of Appeals brought this policy to a virtual halt in 2004. Therefore margins would surely have been driven down to very low levels by entrants using the incumbents' platforms. AT&T and MCI may have enjoyed a fleeting advantage because of their established brand names, but simply attaching the resale of local service to their long-distance services, which are now experiencing a staggering revenue decline, was not likely to save them (see chapter 6). They need new services—new ideas—but these are not likely to be developed on someone else's network.

Welfare Gains from Local Competition

Since 1996 entrants have expended enormous capital resources to wrest 18 percent of access lines from incumbents, with an uncertain effect on economic welfare. The rather limited evidence reviewed earlier in this chapter suggests little difference between entrants' services and those of the incumbents or even in the prices charged to subscribers. Nevertheless, it is difficult to believe that subscribers would have shifted from incumbents to the entrants unless they were offered lower rates for the same service or higher service quality at the same rates.

The data on local telephone rates reflect very little change since 1996 despite the advent of competition. The incumbents' local rates are still regulated by the states and did not move perceptibly from 1996 to 2002.[61] Between 1999 and 2002, the FCC raised the subscriber line charges attached to residential bills by approximately $2.00 a line in order to reduce access charges on interstate long-distance calls.[62] Net of this increase, the average urban rate for incumbents' local residential service in 2002 was $21.32. In 1995 the average rate was $20.01.[63] Regulated local residential rates did not decline in nominal terms over these six years; in fact, they increased by about 44 percent of the increase in the average consumer price index.[64]

Similarly, the Bureau of Labor Statistics producer price index and consumer price index for local telephone service show little movement, even in constant dollars (see figure 4-7). This is not so surprising given that entrants had captured about 10 percent of residential and small business lines by 2002 and that the incumbents' rates have not moved in nominal dollars.[65] Even if the entrants were offering local rates that were, say, 15 to 20 percent lower than the incumbents' rates, they would only have reduced the *average* local rate by 1.5 to 2.0 percent by 2002, assuming no corresponding decline in incumbent rates. The effect of this decline would only be sufficient to offset 16 to 21 percent of the effect of the FCC's increase in subscriber line charges.

The FCC reported that competitive local carriers received $17.7 billion in end-user revenues in 2003.[66] It also reported that the CLECs had 16.7 million small business and residential lines as of June 30, 2003. Given that the average residential subscriber spent $441 in 2003 on services from local carriers, the CLECs' small business and residential revenues may be estimated at $7.4 billion, leaving $10.3 billion in end-user revenues from medium and large business ("enterprise") customers.[67] Assuming that the residential and small business customers received discounts of 15 percent from the incumbents' rates, these customers saved $1.3 billion in 2003.[68] Given that these lines are likely to be delivered mostly through the UNE-P, there is no innovation in such a service, and residential and small business demand is extremely price inelastic. As a result, there is no increase in output to consumers. If the competitors' vari-

Figure 4-7. *Producer and Consumer Price Indexes for Local Telephone Service, Constant Dollars, 1996–2003*

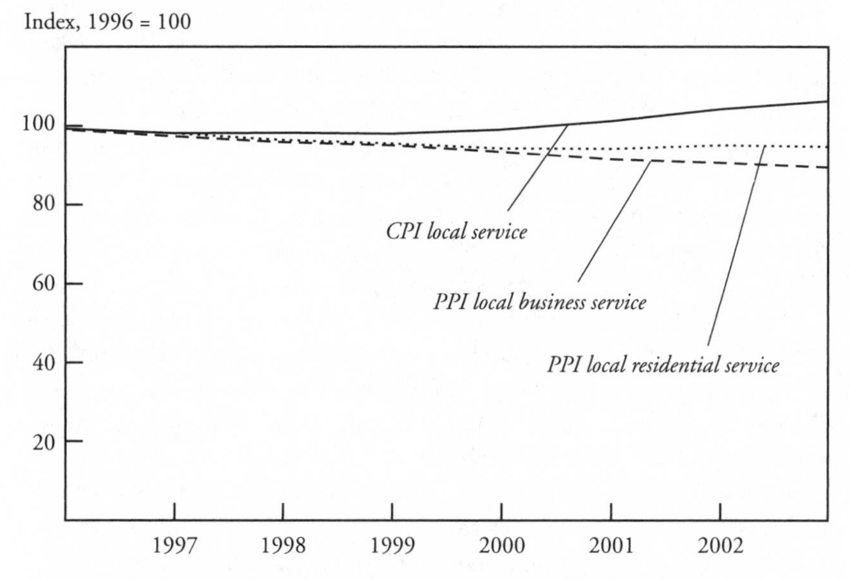

Source: Bureau of Labor Statistics, *Consumer Price Index and Producer Price Index*; Bureau of Economic Analysis, *Gross Domestic Product* (chain price deflator).

able costs are no greater than the incumbents' avoided costs, an extremely unlikely assumption, the amount of noncapital resources devoted to these local services is unchanged, and there is no net welfare gain or loss to the economy other than the investment in duplicative capital facilities. Moreover, the 15 percent decline in price is entirely a transfer from the incumbents to consumers.

The CLEC competition in the larger business market is focused on special access services and various private line services, including high-speed lines. The incumbent Bell carriers reported $14.4 billion in special access revenues and $1.9 billion in private line revenues in 2003, but the special access revenues include DSL revenues.[69] Assuming that the average DSL revenue was $480, the 6.5 million Bell DSL subscribers in mid-2003 generated $3.1 billion in revenues, which must be deducted from the Bell companies' total of $16.3 billion in special access and private line revenues. This leaves $13.2 billion in revenues to be added to the $10.3 billion of larger-company CLEC revenues, or $23.5 billion. It has been estimated that Bell company special access revenues per line fell by about 30 percent in constant dollars between 1996 and 2003.[70] Assuming, quite generously, that this entire decline is due to increased competition from CLECs and that the price elasticity of demand for special access and private lines is -0.5, the gain to business consumers is equal to $9.2 billion, but $8.4 billion of this gain is simply a transfer from the incumbents to business customers.[71] Thus the net welfare benefits in the business market are $0.8 billion. Since there are no net welfare benefits in the residential market, the total net welfare benefits from local competition may be estimated at $0.8 billion in 2003. The total consumer welfare gain is much larger, of course, totaling $10.5 billion, but $9.7 billion is simply a transfer from producers to consumers.

If the CLECs served their business and residential customers at the same variable costs as the ILECs (surely an extremely generous assumption), the cost of conveying these benefits would only be the annual capital costs of the entrants.[72] Between 1996 and 2003, the publicly traded entrants—excluding AT&T, WorldCom, and the cable television companies—invested $35.8 billion in capital facilities.[73] A conservative estimate of total entrants' capital spending over this period would be $55 billion; the entrants' trade association claims that $75 billion was invested between 1996 and 2003.[74] Even with only a 15 percent capital charge to cover the before-tax cost of capital and depreciation, the annual capital charges required to amortize this cumulative investment would have been in excess of $8 billion.[75] Thus the entrants would have incurred at least $8 billion in annual capital charges to generate a net welfare gain of just $0.8 billion in 2002.

These calculations ignore the considerable expenditures made by the entrants on marketing and general administration, including executive salaries. According to financial analysts, in 2000 the largest entrants were spending approximately one-half of their revenues on sales, general, and administrative (SG&A) expenses.[76] By comparison, the SG&A expenses of the large incumbent GTE were only 17 percent of revenues in 2000.[77] The high entry costs are understandable since most of the new entrants were still in a start-up phase and had to spend large sums on marketing to shift customers to a service that was little different from what they already had. But these higher costs must also be deducted from any potential gains of the new competition. Indeed, the additional SG&A by itself swamps any conceivable estimate of the net benefit to the economy.[78]

By 2000 the new entrants were accounting for almost 12 percent of employment in the fixed-wire sector.[79] At the same time, they had a much smaller share of industry revenues and added even less to national output

Figure 4-8. *Labor Productivity growth in Wired Telecommunications, with and without the CLECs, 1987–2002*

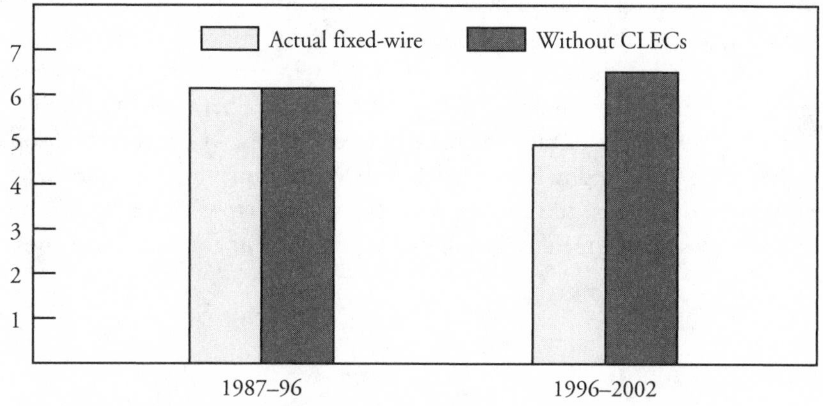

Source: Bureau of Labor Statistics and author's calculation.

because much of their revenue derived simply from reselling the incumbents' services. As a result, the CLECs were a substantial drag on industry productivity growth between 1996 and 2002. To estimate the extent of this drag, I adjust the Bureau of Labor Statistics productivity figures for CLEC output and labor inputs.[80] As figure 4-8 shows, without the CLECs the fixed-wire sector's productivity growth would have accelerated, but at a lesser rate than that experienced by the wireless carriers (see chapter 3).

One might argue that the discounted value of *future* gains could offset the $55 billion in additional capital expenditures made by entrants in 1996–2003, but given the collapse of many of these companies, it seems unlikely that entrant revenues will grow very much in the next few years. Surely these revenues would not defray both the capital charges and the additional SG&A incurred by the CLECs. In short, most of the cost of the 1996–2003 exercise in promoting local entry must at this point be written off as a failed experiment. This is not to say that competition will not emerge or has not emerged (see chapter 7). Rather, the competitors induced into the marketplace through 2003 by regulatory incentives designed to encourage resale in one form or another have not generated benefits that can justify their huge investments in facilities, start-up costs, and marketing expense.[81]

Conclusion

The FCC no doubt attempted to implement the 1996 act with the best of intentions, hoping that its liberal "interconnection" policy would encourage sustainable entry. In fact, a large number of entrants did appear, investing at least $55 billion in capital facilities. Eight years later, few of these entrants remain viable. Entry has not reduced subscriber rates measurably, nor has it provided a notable increase in new services.[82] Local competition seems to be settling down to a battle between the incumbents, the cable television companies, and the wireless carriers. In chapter 5, I examine the effect of these failed regulatory policies on the incumbent Bell companies.

5 Effect of the 1996 Act on Incumbent Local Companies

There can be little doubt that the incumbent telephone companies have lost subscribers to the new entrants since 1996. Between 1996 and 2000, a period of vibrant economic expansion, these losses occurred against a backdrop of subscriber line growth and a generally buoyant stock market. Now, however, the total number of access lines in the country is declining as residences reduce their use of second lines and some simply rely on wireless service as their only telephone service.[1] More important, the equity markets in general have soured on telecommunications. In this environment, the incumbents' loss of retail (end-user) lines to entrants is even more painful than if it had occurred when total line counts were increasing.

Between 1984 and 1996, total access lines increased about 3 percent a year (table 5-1). Virtually all of this growth was registered by the incumbent local companies, principally the Bell companies and GTE.[2] The growth in access lines accelerated after 1996, fed by the demand for second lines due to the Internet, but then slowed after 2000 as broadband services began to replace ordinary voice-grade connections to the Internet and the economy turned sluggish and even became negative after 2001. At the same time, the competitive local exchange carriers (CLECs) trebled their number of access lines, replacing the Bell companies' and other incumbents' end-user lines with services offered over the incumbents' wholesale lines and over their own lines. In the face of these three forces, incumbents

Table 5-1. *Switched End-User Access Lines, Selected Years, 1984–2003*
Millions at Year-End[a]

Year	*Total lines*	*Incumbents*	*Competitive carriers*
1984	113.8		
1994	151.5		
1999	189.5	181.3	8.2
2000	192.6	177.6	14.9
2001	191.7	172.0	19.7
2002	187.5	162.7	24.8
2003	181.4	151.8	29.6

Source: FCC, *Trends in Telephone Service* (May 2004), table 7.1; FCC, *Local Telephone Competition: Status as of December 31, 2003* (June 2004), table 1.

a. Total lines for 1984 and 1994 are not strictly comparable with data for later years.

suffered a decline of 16 percent in their end-user lines between the end of 1999 and 2003.

Bell Company Equity Values

Since 1996 the large Bell company incumbents have seen their equity values rise and fall with the general stock market (figure 5-1). They underperformed the S&P 500 in early 2000 in response to an adverse Supreme Court ruling that affirmed the Federal Communications Commission's (FCC's) right to establish pricing rules for unbundled elements, but they rebounded in late 2000 and 2001.[3] In the spring of 2002, they began to lag behind the S&P 500 once again in the wake of additional adverse legal rulings concerning broadband regulation and evidence that local wireline services were losing traffic to wireless operators. In February 2003, the FCC's adverse ruling on UNE-Ps drove Bell company equities down, and though it was reversed by the U.S. Court of Appeals early the next year, Bell company stocks had still not recovered by the end of June 2004. Nevertheless, of the U.S. telecom carriers, the Bell companies have been the least affected by the telecom bubble and the subsequent meltdown that began in 2000.[4] The reasons for this result are to be found in the post-1984 regulatory history of the telecommunications industry.

When divested from AT&T in 1984, the Bell operating companies were sequestered in the local-exchange business by the AT&T antitrust decree.

Figure 5-1. *RBOC Equities versus Standard & Poor's 500*[a]

Index, February 1996 = 100

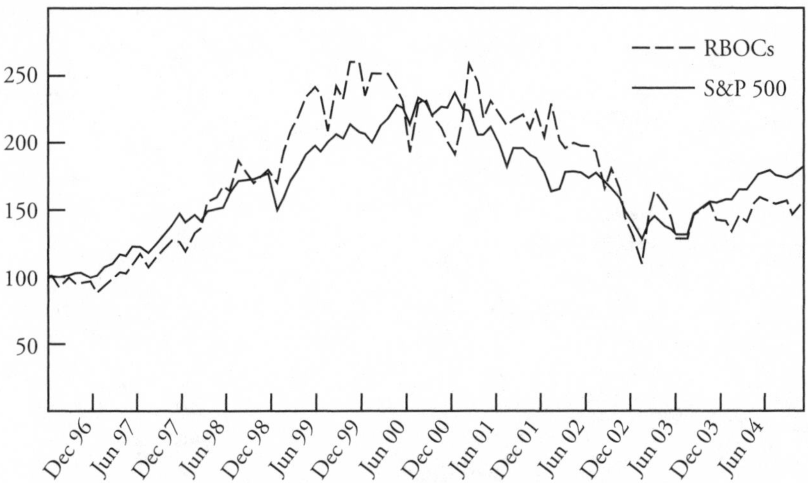

Source: Author's calculation from data obtained from www.finance.yahoo.com.

a. RBOCs in this index are Bell South, SBC, and Verizon only. US West was acquired by Qwest, a long-distance company, in 2000.

For twelve years, they attempted to gain entry into the more dynamic parts of the telecommunications business, with only modest success.[5] Indeed, the 1996 act was largely the result of the frustration that the Bell companies felt in trying to free themselves from the shackles of the 1982 AT&T decree. They negotiated their potential freedom by agreeing to the regulatory requirements built into the 1996 act in return for eventual entry into long-distance services, including those required to function in the Internet economy.

Six years after the passage of the 1996 act, the Bell companies had still not attained the freedom they had bargained for in Congress. By the middle of 2002, they had obtained approval for long-distance entry in states that accounted for only about one-third of the country's access lines (see table 5-2). The slow pace of state and FCC approval had kept them from attempting to develop long-distance networks, Internet backbones, and related Internet services. Moreover, regulators had refused to relieve them of regulating the facilities required to deliver broadband digital subscriber line (DSL) services. By early 2004, however, they had shed some of these shackles. Between June 2002 and December 2003, the Bell companies received

Table 5-2. *Regulatory Approval of Bell Company Entry into in-Region InterLATA Services*

Company	*State*	*Date of approval*	*Switched access lines, 12/31/01 (thousands)*
Verizon	New York	12/99	12,591
SBC	Texas	6/00	10,399
SBC	Kansas, Oklahoma	1/01	2,825
Verizon	Massachusetts	4/01	4,374
Verizon	Connecticut	7/01	2,390
Verizon	Pennsylvania	9/01	7,929
SBC	Arkansas, Missouri	11/01	4,050
Verizon	Rhode Island	2/02	649
Verizon	Vermont	4/02	369
BellSouth	Georgia, Louisiana	5/02	6,437
Verizon	Maine, New Jersey	6/02	7,835
BellSouth	Alabama, Kentucky, Mississippi, North Carolina, South Carolina	9/02	15,724
Verizon	Delaware, New Hampshire	9/02	1,389
Verizon	Virginia	10/02	4,696
BellSouth	Florida, Tennessee	12/02	13,682
SBC	California	12/02	22,285
Qwest	Colorado, Idaho, Iowa, Montana, Nebraska, North Dakota, Utah, Washington, Wyoming	12/02	10,841
Verizon	D.C., Maryland, West Virginia	3/03	5,795
SBC	Nevada	4/03	1,309
Qwest	New Mexico, Oregon, South Dakota	4/03	3,119
Qwest	Minnesota	6/03	2,230
SBC	Michigan	9/03	5,608
SBC	Illinois, Indiana, Ohio, Wisconsin	10/03	19,473
Qwest	Arizona	12/03	2,901
Total approved through June 2002 (percent of U.S. total, excluding Alaska and Hawaii)			59,848 (36.3)

Source: FCC, *RBOC Applications to Provide In-region, InterLATA Services under § 271* (www.fcc.gov/Bureaus/Common_Carrier/in-region_applications/); FCC, *Statistics of Communications Common Carriers,* 2001–02 edition.

Table 5-3. *Measures of Equity Risk in Various Industries*

Sector	*Range of betas*
Bell companies	0.8–1.0
Large trucking companies	0.3–1.3
Four major railroads	0.4–0.6
Major airlines	1.0–2.8
CLECs	1.0–4.5
Long-distance companies	0.7–1.4

Source: http://quote.yahoo.com (July 11, 2004).

regulatory approval to begin offering interLATA long-distance services in all of the remaining states, and in February 2003 they received some relief from the FCC on the requirements for unbundling and line sharing for broadband (see chapter 2). Finally, in early 2004, the U.S. Court of Appeals overturned the FCC's attempt to retain the unbundling regime that permits competitors to lease the entire UNE platform.[6]

Before 2002 the only major segment of domestic telecommunications that the Bell companies had been able to enter without the restraint of regulation was commercial wireless service (cellular). These wireless operations have accounted for between 15 and 30 percent of the companies' market value since 1996. Until recently, the remainder had been a no-growth, "public utility" business that was relatively secure, unexciting, and almost devoid of incentives to innovate. The best evidence of this situation is to be found in the financial markets' evaluation of the risk of investing in their equities.

As table 5-3 shows, the Bell companies' equities have beta values of risk somewhat less than 1.0. The higher the beta measure, the greater is the market's view of the risk in holding the company's shares.[7] A beta of 1.0 is average for all traded risky assets. Two firms with beta values close to 1.0 are Ford and General Motors, while airlines have betas of 1.0 to 2.8. The financial markets apparently viewed the Bell companies as generally less risky, or at least they did so until recently. Beta values go up to 4.5 for the largest CLECs, which are failing rapidly, but are in the range of 0.7 to 1.4 for long-distance companies.

The beta estimates in table 5-3 are provided by a commercial source and are apparently estimated from stock-price movements over a considerable period of time. To examine the recent trend in beta estimates more closely

Table 5-4. *Estimates of Equity Betas, 1996–2003*

Company	*1996–97*	*1998–99*	*2000–01*	*2002–03*	*Entire period*
Bell South	1.38	0.31	0.24	0.90	0.58
SBC	1.26	0.71	0.44	1.08	0.80
Verizon	1.15	0.62	0.43	0.77	0.69
Qwest	n.a.	1.42	0.79	1.20	1.01
AT&T	0.80	0.70	1.09	0.77	0.83

Source: Author's estimates from closing prices on http://finance.yahoo.com.
n.a. Not available.

(table 5-4), I calculated each company's beta from weekly closing prices for two-year periods between 1996 and 2003 and for the entire eight-year period. Note the relatively high estimates for Bell company equities about the time the telecom act was passed and in the early stages of implementation, in 1996–97. Clearly, this was a period of great uncertainty for these companies.

For the next four years, however, the Bell companies' equity betas fell sharply, as they were now confined by the act to their intrastate and local markets and were losing few lines to competitors. Regulation had apparently succeeded in bottling up the resources of a large part of the telecommunications sector and was treating it like a staid trucking company or railroad. The Bell companies were prevented from making the mistakes of their unconstrained fellow telecom companies, but the country was deprived of the potential innovation that might have come from this reservoir of expertise in telecommunications.[8]

Once the Bell companies were allowed to offer interstate services and were saddled with ever lower rates for their wholesale services, the risk of holding their stocks rose markedly.[9] Although not quite back to 1996–97 levels, their estimated betas are now about 1.0 on average. The uncertainty over the FCC's unbundling rules, the loss of lines to competitors, and the more competitive era unleashed by Bell company entry into interLATA long-distance markets has made the Bell stocks more risky to hold.[10] In response, the equity markets did not raise AT&T's beta very much, which is surprising.

The overall effect of the new regulatory order on the Bell companies has thus been rather mixed. On the one hand, the companies were kept out of interLATA markets until recently and have been required to provide wholesale facilities to their rivals, thereby having less incentive to invest. On the other hand, they saw their equity prices track the entire stock market with betas of 0.3 to 0.7 for several years. As a result, the market value of the Bells' wireline assets has grown modestly since 1996, whereas the value of their enterprises per access line has not changed very much. At the end of 1996, the Bell company nonwireless assets were valued at about $1,940 per switched access line; by the end of 2003, they were valued at about $2,170 per line, an increase of 12 percent during a period in which the overall stock market grew by 50 percent.[11] Given the surge in capital expenditures by these companies in 1998–2001, as they modernized their facilities to deploy broadband DSL services, this modest growth could hardly be reassuring to the Bell companies, even if one interprets the impact of the 1996 act on the incumbents as relatively benign.

Threats to Bell Company Revenues

Despite very little change in local rates since 1996 and little growth in retail access lines, the incumbent local exchange carriers (ILECs) recorded wireline revenue growth of approximately 20 percent between 1996 and 2001.[12] A large share of this growth came from second lines required for dial-up Internet services.

The slow, steady growth of the incumbent Bell companies' wireline operations has been threatened from at least three directions. First, the loss of end-user lines (subscribers) to the new CLECs, mostly in the form of leased wholesale lines or the entire UNE-P platform, has reduced the Bell companies' revenues by far more than their costs. The average regulated wholesale price of leasing the entire UNE platform declined steadily through 2003 owing to pressure on regulators to provide the entrants with greater potential operating margins. By November 2002, these UNE platform rates had fallen to about 40 percent of the incumbents' retail revenues per line (see table 4-1).[13] These low rates for essentially reselling the incumbents' services over their entire platform had attracted entrants and contributed 8 percentage points of the incumbents' 16 percent loss of retail lines by the end of 2003 (see figure 4-1). The replacement of retail revenues with wholesale revenues, in turn, reduced Bell company revenues by about 5 percent. Since the Bell companies' avoided costs of customer service and marketing are only about 10 percent of their total costs, their cost savings from ceding the retailing function pale in comparison with their revenue losses from leasing the UNE platform, as shown in the last two columns of

Table 5-5. *Effects of CLEC Use of UNE Platform on the Bell Companies, 2002*

Company	*Average wholesale UNE-P rate (dollars/month)*	*Average retail revenues/line (dollars/month)*	*Lines leased at UNE-P rates*[a]	*Revenue loss from UNE-P (millions of dollars/year)*[a]	*Savings in marketing and customer service (millions of dollars/year)*
BellSouth	23.10	53.69	1,359,000 (6)	499 (3)	95
SBC	16.55	51.23	3,851,000 (7)	1,603 (5)	253
Verizon	19.40	42.49	2,574,000 (6)	713 (4)	112
Qwest	22.94	51.10	n.a.	n.a.	n.a.

Source: Anna Maria Kovacs, Kristin Burns, and Gregory S. Vitale, *The Status of 271 and UNE-Platform in the Regional Bells' Territories* (Commerce Capital Markets, November 8, 2002).

n.a. Not available.

a. Figures in parentheses represent percent of total.

table 5-5. These data show that the three largest Bell companies' losses from the UNE-P had reached $2.4 billion by mid-2002. By mid-2004 the number of UNE-P lines had more than doubled, suggesting a loss of $5 billion a year for the Bell companies. However, the federal appeals court reversal of the FCC's unbundling rules in early 2004 has probably brought an end to this source of erosion of Bell revenues.

Second, all wire-based local telephone carriers are likely to suffer increasing losses in lines to wireless carriers, cable television companies, and broadband services. The decline in wireless rates and the national pricing plans first introduced in 1998–99 have already led some households to drop their wireline service altogether. Broadband services, whether in the form of DSL, cable modem services, or wireless services, reduce the need for residences to subscribe to second lines. Since 1997 the number of U.S. fixed lines has declined by 11 million, from 192.5 million to 181.4 million.[14]

Equally important, the cable television companies are steadily expanding their telephone service over their own coaxial cable networks. As cable modem service spreads, they will begin to offer their subscribers voice over Internet protocol (VoIP) services at very low rates.[15] The potential advantages of VoIP are substantially enhanced by current regulatory policies. If the VoIP services do not have to pay federal universal service charges, which are now 9 percent of interstate and international revenues, and can avoid regulated interstate and intrastate switched access charges on long distance, they will have a sizable cost advantage over traditional wire-based local and long-distance carriers. As a result, the growth of VoIP threatens the incumbent local carriers with revenue losses, not only from declining numbers of local access lines, but also from the loss of switched access charge revenues that are used to cross-subsidize local service. Even as recently as 2003, the Bell companies realized more than 7 percent of their local revenues from these access charges.[16] State regulators, in particular, have been reluctant to allow these access charges to decline despite threats from wireless and VoIP, because they are unwilling to allow incumbents to raise fixed monthly rates to defray the potential revenue losses.[17]

The likely impact of wireless and cable television competition on incumbent telephone companies may be deduced from the proliferation of bundled pricing plans offered by both. Table 5-6 shows a selection of the plans available in mid-2004. For $60 a month, subscribers can now get 800 to 1,000 minutes of local and long-distance minutes during peak calling hours and unlimited minutes on weekends or on nights and weekends; for $100, they can purchase up to 2,500 peak minutes plus unlimited off-peak minutes. In some cases, these plans even allow free "roaming" to areas outside the subscriber's home market.

These wireless plans are competitive with the incumbents' wireline services for a substantial number of customers. The average bill for residential wireline service, including long-distance, is now about $47 a month.[18] However, a substantial share of subscribers spend much more, primarily because of extensive long-distance calling. If this calling is concentrated during evening or weekend hours, the subscriber can save money by dropping the wireline service and subscribing to one of the plans shown in table 5-6. Unless the subscriber makes large numbers of zero-priced, local peak-hour calls from his or her residential telephone, using a wireless phone for local calls is likely to be less expensive than the $35 a month for local wireline service, including optional services. Given the continuing decline in wireless rates and the relatively constant regulated incumbent telephone company local rates, wireless/wireline substitution will surely grow (see chapter 7).

At the same time, cable companies like Cox, Comcast, and Cablevision are offering a variety of local and long-distance plans that are competitive with incumbents' services. For instance, Cox offers a number of local/long-distance calling plans for $25 to $39.95 a month.[19] By late 2002 it

Table 5-6. *Wireless and Cable Telephone Pricing Plans, Late 2004*

Company	*Service*	*Monthly rate (dollars)*
Wireless		
T-Mobile	1,000 anytime, unlimited night and weekend minutes	59.99
	2,500 anytime, unlimited night and weekend minutes	99.99
Cingular	850 anytime, unlimited night and weekend minutes	59.99
	2,000 anytime, unlimited night and weekend minutes	99.99
Verizon	800 anytime, unlimited night and weekend minutes	59.99
	2,000 anytime, unlimited night and weekend minutes	99.99
AT&T	800 anytime, unlimited night and weekend minutes	59.99
	1,800 anytime, unlimited night and weekend minutes	99.99
Sprint	700 anytime, unlimited night and weekend minutes	50.00
	2,000 anytime, unlimited night and weekend minutes	100.00
Cable		
Cablevision	Unlimited local and long distance over VoIP in New York	34.95
Comcast	5,000 local and long-distance minutes/month in New England	49.00
Cox	Local service plus 500 minutes long fistance in California, Connecticut, and Rhode Island or 90 minutes long distance in other states	25.00 to 39.95

Sources: Various company web sites and news releases, July–October 2004.

had already attracted as many as 30 percent of its cable subscribers to its telephone service in its most mature markets.[20] Given the cable companies' overwhelming share of residential broadband connections (see chapter 8), they are now beginning to offer even lower-priced Internet telephony service.[21]

Third, until recently, the Bell companies were losing the broadband race to their unregulated cable television rivals. Because of this, their opportunity for growth had appeared to be diminishing, a fact that they recognized by reducing their capital expenditure budgets in 2002–03 to levels below those of 1995–96 (see chapter 8 and the next section). The FCC's 2003 decision to discontinue line sharing for broadband services may have stimulated Bell company deployment of DSL and is apparently allowing them to narrow the gap with cable modem service (see chapter 8).

Together, these forces have eroded Bell company revenues. All incumbent local companies' revenues fell by 2.5 percent in 2002, from $117.9 billion to $115.0 billion, and a further 4.5 percent in 2003.[22] Furthermore, analysts were predicting that the Bell companies' fixed-line revenues would decline slowly for the next few years as increases in DSL revenues failed to offset the revenue losses from the rapid growth in UNE-P lines.[23] Given the uncertainty surrounding FCC and state regulation of wholesale prices, services, and VoIP, these projections are subject to substantial error, but they are obviously reflected in the performance of Bell company equities, as figure 5-1 shows.

Capital Expenditures

The pessimism surrounding the Bell companies has obviously affected their capital expenditure budgets. After substantially increasing capital spending between 1998 and 2000 to prepare for the broadband revolution (see figure 5-2), incumbent carriers trimmed their budgets in 2002–03. The problem was not "excess capacity"—so common elsewhere in the telecommunications sector—since the Bell companies were blocked from interLATA services between 1996 and 2000. If anything, they lack the capital facilities to deliver the new DSL services (see chapter 8). Rather, the economic and regulatory environment forced incumbent local carriers, including the Bell companies, to reduce their capital spending more than their major rivals in local communications markets, namely, the cable companies and wireless carriers. Of course, CLEC investment has collapsed because none of these companies has found a sound business mode (for an analysis of entrants, see chapter 4).

How far the wholesale unbundling regime affected Bell companies' incentives to invest is the subject of lively debate. Many claim that the low wholesale UNE and UNE-P rates and the requirement of line sharing reduce the incumbent local companies' incentives to invest because they are forced to share the fruits of their investments with rivals at low, total-element, long-run incremental cost (TELRIC). Others argue that low wholesale rates prod the Bell companies to invest more so as to fend off the competition unleashed by low UNE and UNE-P rates. In support of this suggestion, Robert Willig and his associates have found that Bell company capital expenditures across forty-eight states in 1996–2000 and 1996–2001 varied inversely with UNE-P rates in *June 2002*, holding other influences constant.[24] While it is possible that lower wholesale rates induce the Bell companies to invest more in their networks to cope with prospective increases in competition, such a response seems unlikely. Why would a

Figure 5-2. *Capital Expenditures by U.S. Local Telecom Operators, 1996–2003*

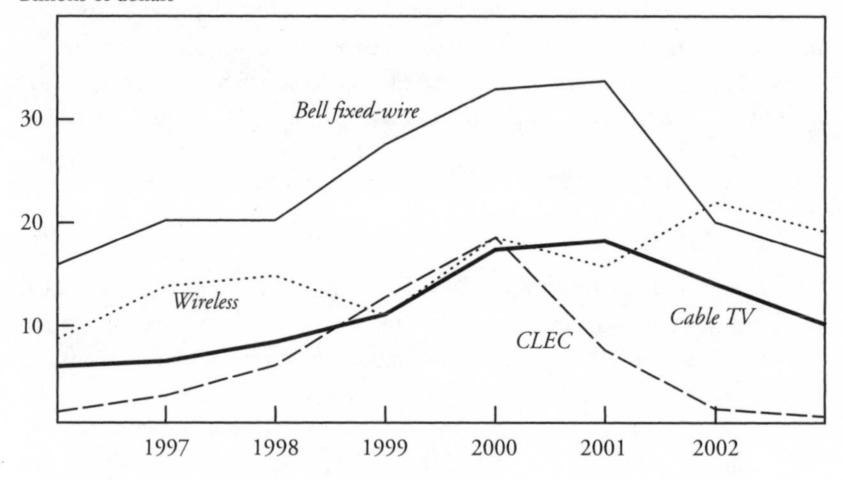

Source: Author's estimates from company annual reports to the SEC (www.sec.gov/cgi-bin/srch-edgar); and CTIA, *Semi-annual Wireless Survey* (December 2003).

company invest more in facilities that it has to lease at lower rates? Furthermore, one cannot assume that investment responds to UNE-P rates if those rates are measured for a period *after* the capital expenditures take place. One would have to show that *subsequent* investment expenditures fall or rise with differences in UNE-P rates.

Given that twenty-one states reduced UNE rates between early 2001 and July 2002, it is useful to consider how the Bell companies adjusted capital spending in these states compared with other states.[25] Between 1996 and 2000 (that is, before the reductions), the Bell companies' capital spending in both groups of states grew at the same rate. In 2001, however, capital spending grew more rapidly in the states that did not change UNE rates than in those that lowered their UNE rates.[26] In 2002 Bell company capital spending fell dramatically in response to the economy and continuing regulatory uncertainty, but it declined more in states that were reducing UNE rates than in those that did not (see figure 5-3). This is hardly conclusive evidence of the adverse investment incentive effects of low UNE rates, but at least it correlates changes in investment spending with regulatory changes occurring before or during the period that the expenditures are occurring.

Figure 5-3. *Capital Expenditures by Bell Companies: States with UNE Rate Reductions in 1996–2002 versus States with No UNE Changes*

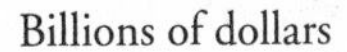

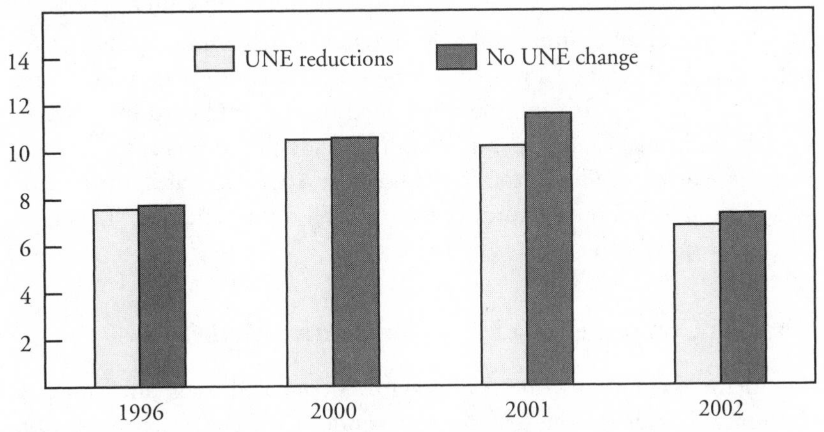

Source: FCC, Armis data (http://www.fcc.gov/wcb/armis/); Billy Jack Gregg, *A Survey of Network Unbundled Element Prices in the United States* (National Regulatory Research Institute), periodic issues.

When I attempted to replicate the regression analysis of Willig and his associates using various published estimates of UNE-P rates for 2000–03, I found that the inverse relationship between capital spending and the wholesale (UNE-P) rates they identified is much stronger in the 1996–99 period than in 2000–03 (see the appendix for details). This result suggests that Bell companies' investment decisions for 1996–99 responded more strongly to 2000–03 UNE-P rates than did their 2000–03 decisions, which is simply not plausible. A more reasonable inference would be that greater capital spending in 1996–99 led to lower wholesale rates in 2000-03 as regulators insisted that the Bell companies pass on the lower costs induced by such spending to their rivals in the form of lower wholesale rates. Indeed, a simple regression of the UNE-P rate in 2002 on the FCC's measure of costs, the state regulatory variables, and the Bell company's 1996–99 capital spending in that state provides a statistically significant negative coefficient on the 1996–99 capital spending.

If Bell companies adjust their investment plans to variations in the UNE-P rate because they think low UNE-P rates will lead to more competition, this effect could be modeled more directly by using the actual

CLEC share of lines rather than one of the contributors to it, the UNE-P rate. When I use this measure of competition as a substitute for the various estimates of the UNE-P rate, I find that competition has no statistically significant effect on Bell company capital spending in 1996–99 or 2000–03.

Why had competition not yet stimulated a Bell company investment response by 2002? The answer must be that regulatory uncertainty offset the influence of rivalry and the threatened loss of retail customers. Moreover, if additional capital spending simply made the company's infrastructure more attractive for entrants to lease at low UNE-P rates, the benefits of additional capital expenditures would surely be attenuated. This latter threat has now been removed by the courts.

The Bell Companies in a More Competitive Market

The Bell companies had been trapped in a regulated, stagnant, and even declining market for the first six years under the new act. Banned from interLATA services in their own regions, they could not participate in long-distance services or in services delivered over the Internet. As they wrestled free from the vertical quarantine imposed by the 1996 act, the Bell companies became much more aggressive in competing with their new cable, wireless, and (to a lesser extent) CLEC rivals, offering a variety of bundled plans and lower-priced DSL services (see table 5-7).

Whether this more aggressive pricing strategy has succeeded in stemming the Bell companies' losses in revenues is far from clear. These companies certainly have garnered a substantial share of residential long-distance customers and are now able to compete for the larger business customers. By early 2004 the courts had all but stopped the growth of the UNE platform, and the FCC had reversed its position on mandating line sharing for broadband connections. Nevertheless, the prospects for Bell company revenue growth from their fixed-wire networks are not very good. At best, they are likely to maintain their current level of nominal revenues. At worst, if VoIP begins to grow rapidly, revenues may continue to decline.

The Bell Companies versus Other Incumbents

When the Bell companies were divested from AT&T in 1984, the incumbents consisted of the Bell group, a number of other independent telephone companies, and the partly owned Cincinnati Bell and Southern

New England Telephone Company. These latter companies were not constrained by the line-of-business restrictions in the AT&T consent decree or by the subsequent competitive "checklist" requirements in Section 271 of the 1996 Telecommunications Act. By the late 1990s, there were only three large publicly listed independents left: Cincinnati Bell, ALLTEL, and Citizens. The others had become part of much larger companies or were too small to be listed on the national stock exchanges.[27]

These three independent companies were free to pursue expansion strategies in the new telecom world unlocked by the 1996 act since they were not limited to providing intraLATA services. In fact, two of them, ALLTEL and Citizens, remained local carriers that did not change their modus operandi significantly, but Cincinnati Bell changed its name to Broadwing, acquired a new national carrier, IXC, and launched an ambitious expansion strategy. In 2000–01 Broadwing spent more than $1 billion on fiber-optic transmission facilities, despite the fact that its national, broadband transmission division had annual revenues of just $860 million in those years.[28] Its total capital expenditure budget had only been $143 million in 1998, but it spent $844 million on capital facilities in 2000 alone. Like many of the long-distance companies and CLECs, Broadwing plunged into the new Internet age with abandon. The results were predictable.

A comparison of the stock market performance of the Bell companies and the three large, unconstrained ILECs reveals a surge in Broadwing's (Cincinnati Bell's) common stock value in 1999 and then its collapse in 2000–02 alongside more stable prices for the other ILECs, ALLTEL and Citizens (figure 5-4). ALLTEL and the Bell companies recorded a similar performance, but Citizens—which owns mostly low-growth rural companies—saw its stock price barely change at all during this tumultuous period. It is notable that an incumbent can be almost unaffected by the 1996 act, particularly if it operates in areas where there is little entrant activity.

Outlook for the Incumbent Carriers

Despite the regulatory constraints under which they have operated since 1996, or even since 1984, the Bell companies are still relatively healthy. At the end of 1996, the predecessors to the current Bell companies had a combined market capitalization of $231 billion. By the end of 2003, the four companies had a combined market cap of $241 billion.[29] This is obviously

Table 5-7. *Bundled Pricing Plans Offered by the Bell Companies, July 2004*

Name	*Bundled services*	*Monthly price (dollars)*
Verizon		
Verizon Freedom	Unlimited local and long-distance calling, other features	49.95–59.95
Verizon Freedom with DSL	Unlimited local and long-distance calling, unlimited DSL	Varies; 84.90 in Massachusetts
Verizon Freedom with Internet	Unlimited local and long-distance calling, choice of internet dial-up plan	Varies; 79.95 in Connecticut
Verizon Freedom with Wireless	Unlimited local and long-distance calling, choice of national wireless plan	Varies; 86.44 in Massachusetts
Verizon Freedom All	Unlimited local and long-distance calling, unlimited DSL, choice of national wireless plan	Varies; 116.39 in Massachusetts
SBC		
ALLDISTANCE Connections	Local, long-distance, and other features	48.95
Total Connections DSL	Local, long-distance, DSL, and other features	75.90
Total Connections DIAL	Local, long-distance, internet, and other features	61.90
Bell South		
Value Answers Premier	Local, long-distance, and other features; unlimited minutes	54.99–59.99

DSL Bundle	Local, long-distance, and high-speed Internet	Fixed rate: 78.90–82.90 + 0.05/minute; unlimited rate: 94.94–99.94
Wireless Bundle	Local, long-distance, wireless, and other features	Fixed rate: 68.94–72.94 + 0.05/minute; unlimited rate: 84.98–89.98
Dial + Wireless Bundle	Local, long-distance, Internet, wireless, and other features	Fixed rate: 78.89–82.89 + 0.05/minute; unlimited rate: 89.93–94.93
DSL + Wireless Bundle	Local, long-distance, high-speed Internet, wireless, and other features	Fixed rate: 113.89–117.89 +0.05/minute; unlimited rate: 124.93–129.93
Qwest		
Unlimited Long- Distance Plan	Unlimited long-distance with monthly fee (for use with non-Qwest local provider)	30.00 first year; after that, 34.95
Preferred Unlimited	Unlimited local and long-distance calling with Qwest	45.99 first year; after that, 50.99
Qwest Choice DSL	Unlimited local and long-distance calling, unlimited DSL	72.98 first year; after that, 77.98
Qwest Choice Wireless	Unlimited local and long-distance calling, choice of wireless plan	80.98 first year; after that, 85.98
Qwest Choice DSL + Wireless	Unlimited local and long-distance calling, unlimited DSL, choice of national wireless plan	107.97 first year; after that, 112.97

Source: Company websites, July 26, 2004.

Figure 5-4. *RBOC Stock Prices versus Other ILECs*

Index, February 1996 = 100

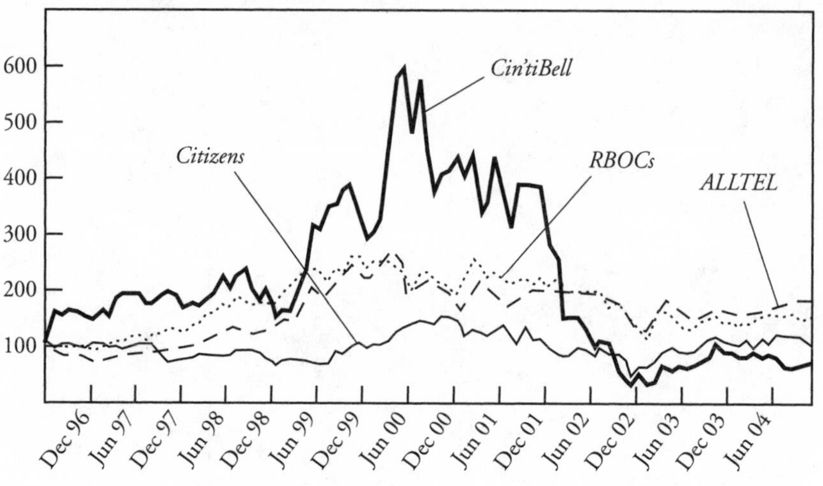

Source: www.finance.yahoo.com.

not a very good performance when the S&P 500 has risen by 50 percent, but it is better than that of most other participants in the world's telecommunications sector over the past few years. The value of many of those incumbents has declined sharply since 2000, after the meteoric rise of the late 1990s (see chapter 9). The U.S. incumbents—other than Broadwing—did not enjoy the sharp increase, but they have not suffered a shattering decline, either. Given their regulatory quarantine, the Bell companies could not engage in the merger spree pursued by the other large U.S. telecommunications carriers, namely, WorldCom, Qwest, AT&T, Global Crossing, and Broadwing.[30] Instead, they consolidated as local carriers, invested in wireless (cellular) service, and slowly built out their broadband capability. With the exception of Qwest, they have now become the most stable firms in the industry.

The Bell companies continue to account for about 85 percent of the country's switched access lines, half of the country's cellular subscribers, and about 30 percent of mass-market broadband Internet lines.[31] Of the other participants in the U.S. telecom sector, only Sprint has a similar breadth of facilities, but its local division is small (only about 8 million lines) and dispersed about the country. No other wireless, long-distance, or new local carrier has the capacity to offer a bundled package of its own services to residences and small businesses.

Although the cable television companies have no wireless assets and only limited local telephony capacity at present, they stand poised to become the most important competitors of the Bell companies because they can offer telephony, broadband Internet service, and video services. The Bell companies have not been able to develop video services over their networks but may begin to do so once they extend fiber-optic lines closer to the home.

As of December 2004, the cable companies are preparing to launch their VoIP services on a wide scale because independent companies, such as Vonage, are beginning to market their own VoIP services to the cable companies' cable modem subscribers. Given the potential cost advantages of VoIP that could be created by current regulatory policy, as already pointed out, these services could grow rapidly in the next few years.

Industry participants other than the cable companies and the incumbent Bell companies lack the facilities and services to offer a bundled product whose dimensions are constantly changing. Carriers with limited facilities, such as the new local entrants (CLECs), are by and large resellers and arbitrageurs that have a limited array of services and cannot develop innovative ones. Long-distance carriers or independent wireless companies may own their own transmission and switching facilities, but—other than Sprint—they cannot match the diverse service offerings of the Bell companies or the cable companies. Without this breadth, marketing costs are likely to be prohibitive. Whereas the Bell and cable companies can advertise and promote a large number of services simultaneously, less diverse carriers may pay just as much to promote a single service. Unless it is an innovative service not available from the Bell companies or the cable companies, the marketing cost per actual enrollee makes the service uneconomical. All in all, in the aftermath of the major shakeout of new local carriers and long-distance companies, the Bell companies are now in a position to offer residences and small businesses an array of communications services, other than video, that only Sprint can match.

6 The Death of Distance and of the Long-Distance Carriers

Before 1970 there were no separate, independent long-distance carriers in the United States. As in other countries, long-distance service was provided by integrated companies that offered local connections and long distance. AT&T, the leading carrier, provided access to more than 80 percent of local access lines and virtually all interstate long-distance service in the country. Entry in long-distance services first appeared in the 1970s, not necessarily because nonintegrated entry made economic sense but because state and federal regulators overpriced long-distance service. If entrants such as MCI or Sprint could gain access to AT&T's and other local companies' lines to originate and terminate calls, they could compete with AT&T in long distance because of these very high regulated rates.

When AT&T tried to deny access to entrants, the antitrust authorities intervened and sued AT&T under the Sherman Act. Ten years later, AT&T was broken up into separate local companies and a long-distance, equipment, and research company. The industry became vertically fragmented as a result of regulatory distortions interacting with antitrust, not through market forces. Once AT&T was separated from its local operating companies, regulators were forced to establish explicit access charges through which the new long-distance companies would pay the local companies to connect their calls. All long-distance carriers were to be provided "equal access"—that is, connections of the same quality at the same rates—for interstate calls. State regulators would not create similar rules for the shorter intrastate calls for years thereafter because the states wanted to allow local companies to continue using excess long-distance charges to defray the costs of local service.[1]

The Major Players

The first new entrant into long distance before the AT&T divestiture was MCI, a carrier established in 1969 with service between St. Louis and Chicago. After the Federal Communications Commission permitted general "special carrier" entry in 1971, a number of other firms began to enter. Among these were the forerunners of Sprint, the third largest carrier.[2] Several others started up in the late 1970s and 1980s, many of which combined into what is now MCI (formerly WorldCom).

By 2000 scores of companies were offering long-distance services, many of them simply resellers of the capacity of the larger carriers. The three largest companies accounted for 70 percent of long-distance revenues (table 6-1). Since that time, however, revenues have begun to decline dramatically. By 2002 many of these companies were in severe difficulty, two (WorldCom and Global Crossing) were in bankruptcy, one (Concert) had ceased operations, and Bell Canada Enterprises had placed one of its carriers (Teleglobe) in bankruptcy. The long-distance competition that began in the 1970s was now working so well that it was driving most carriers into severe financial difficulty and perhaps out of existence altogether.

Long-Distance Arbitrage

For more than a decade after the AT&T divestiture, the new long-distance companies drove down AT&T's market share relentlessly. These companies were attracted by the difference between AT&T's regulated rates and the access charges paid to the local companies to originate and terminate their calls. Given AT&T's large market share and its inability to reduce rates selectively to counter its new rivals because of regulation, AT&T simply watched its share of the market decline, while it pocketed the enormous cash flows created by regulation. The FCC understood that long-distance rates were far too high because it had kept regulated interstate access charges too high.[3] Therefore, it began to drive these access charges down, compensating the local carriers for the lost revenues by adding a rising monthly "subscriber-line charge" to customers' local bills. This was a sensible rate rebalancing policy, and it provided a continuing entry incentive

Table 6-1. *U.S. Long-Distance Revenues*
Millions of dollars

Company	*2000*	*2001*	*2002*	*2003*
AT&T	38,110	33,942	27,531	22,814
WorldCom	22,998	21,259	17,659	16,062
Sprint	9,038	8,424	7,076	6,336
Qwest	3,044	3,180	3,202	2,180
Global Crossing	2,683	2,042	2,098	
Other long-distance and wireless carriers	27,126	24,681	21,242	19,166
Total (excluding incumbent local carriers)	102,999	93,528	78,808	66,558
Incumbent local carriers	6,617	5,772	4,889	12,042
Total	109,615	99,300	83,697	78,600[a]

Source: FCC, *Statistics of Communications Common Carriers, 2003–04* (October 2004), table 1.4.
a. Preliminary.

into long distance because of the lag in the adjustment of retail rates to these lower costs.

Rates and Access Charges

As table 6-2 shows, long-distance rates were far above access charges right after AT&T was broken up. The new competition from MCI, Sprint (or its antecedents), and numerous resellers placed downward pressure on retail rates that AT&T could not counter effectively because it was still regulated until 1995. As access charges declined, from more than 17 cents a minute in 1984 to 6.5 cents a minute in 1995, retail rates followed, but with a lag. As a result, the new carriers and AT&T could reap substantial cash flows from this business. But if access charges were to stabilize, and if AT&T were deregulated, the margins might shrink rapidly in the face of intense competition.

When the 1996 act was passed, the FCC was expected to respond to the requirement that interstate subsidies be made explicit by moving rapidly toward cost-based prices, thereby abandoning the implicit subsidies in the interstate rate structure created by above-cost switched access charges. The act did not force states to rebalance their rates toward cost, but they might have been induced to do so if the FCC were successful in its rebalancing.

Table 6-2. *Average Long-Distance Rates and Carrier Access Charges, 1984–2002*

Dollars/conversation minute

Year	*Interstate and international*	*Interstate (FCC)*	*Interstate (three largest carriers)*	*Interstate switched access charge*
1984	0.32	...	...	0.176
1985	0.31	...	0.304	0.166
1986	0.28	...	0.250	0.146
1987	0.25	...	0.205	0.120
1988	0.23	...	0.195	0.105
1989	0.22	...	0.180	0.092
1990	0.20	...	0.156	0.076
1991	0.20	...	0.146	0.071
1992	0.19	0.149	0.143	0.069
1993	0.19	0.148	...	0.067
1994	0.18	0.135	...	0.068
1995	0.17	0.125	...	0.065
1996	0.16	0.124	...	0.061
1997	0.15	0.108	...	0.056
1998	0.14	0.114	...	0.040
1999	0.14	0.112	...	0.033
2000	0.12	0.088	...	0.024
2001	0.10	0.079	...	0.018
2002	0.09	0.070	...	0.016

Source: FCC, *Trends in Telephone Service* (May 2004); FCC, *Telecommunications Industry Revenues, 2002* (March 2004); Robert E. Hall, *Long Distance: Public Benefits from Increased Competition,* study prepared for MCI (October 1993).

At first, the FCC moved slowly toward further rate rebalancing. However, a consortium of local and long-distance carriers offered the commission a proposal to lower interstate access charges to 1.1 cents per conversation minute for the larger carriers in return for higher monthly charges per line and some absorption of the lost revenues by the large local carriers. Eventually, the FCC approved this "CALLS" proposal, which has now reduced access charges to slightly more than 1 cent a conversation minute.[4]

Once access charges fall to roughly 1 cent a conversation minute, as currently required by the CALLS plan, they obviously cannot be depressed much farther. Indeed, the margins from long-distance service have shriveled now that interstate access charges have stabilized, wireless carriers have begun to offer national calling plans, and the Bell companies have been permitted to enter interstate long distance. With consumer rates often as low as 4 or 5 cents a minute, the margins have shrunk so much that the independent long-distance carriers would be in severe financial difficulty even if they were not losing call volumes at an alarming rate.

Carrier Concentration and Long-Distance Rates

The price data examined here are average revenues per conversation minute as reported by the FCC or the large carriers (see table 6-2). The FCC obtained its data for 1984–91 directly from AT&T, but since then has calculated averages from reported revenue and carrier minutes. Hence any reporting problems will obviously introduce errors. One likely source of bias is the diversion of reported switched access minutes as carriers try to avoid the very high access charges maintained by regulators in the name of "universal service" policy. These access charges are designed to transfer income from subscribers who make large numbers of nonlocal calls to those who do not and to all subscribers in rural areas.[5]

Even after the CALLS plan is fully effective and switched interstate access charges drop to little more than 1 cent a conversation minute, access charges are still likely to be above the long-run incremental cost of originating and terminating calls. Under the current "reciprocal compensation" rule for terminating local calls, for example, carriers charge between 0.2 and 0.3 cents a minute for terminating local calls.[6] In the mid-1990s, interstate switched access charges were twenty to thirty times higher than the current cost-based interconnection rate, providing enormous incentives for long-distance carriers to attempt to circumvent interstate switched access.[7] This circumvention could occur legally through the leasing of dedicated "special access" lines to large customers or illegally by attempting to pass off interstate traffic as local traffic.[8]

Since the shift from switched access charges to special access during the 1990s, at least one carrier, MCI-WorldCom, has been accused of diverting interstate traffic to other channels through which it could avoid switched access charges.[9] If this were true, the FCC's calculation of interstate rates would be biased upward after 1991 because it would have used a downward-biased series on switched access minutes in the denominator of the calculation.

Figure 6-1. *Actual versus Predicted Wireline Interstate Terminating Switched Access Minutes, 1992–2002*

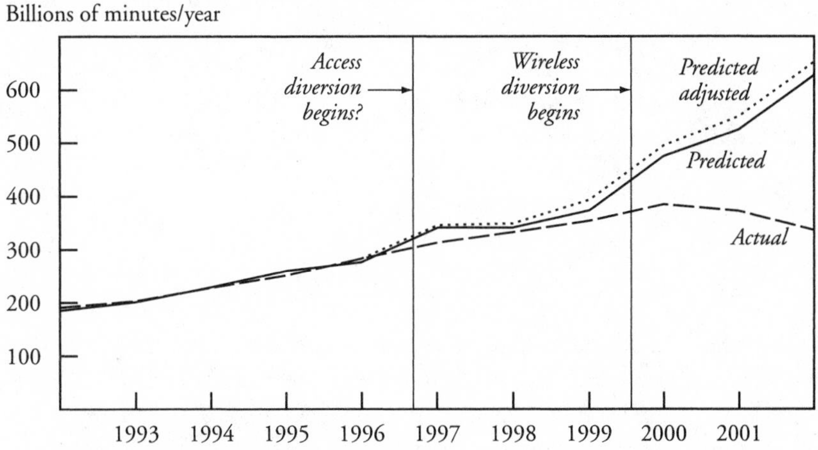

Source: FCC, *Telecommunications Industry Revenue Data, 2002* (March 2004); Bureau of Economic Analysis (GDP); author's calculations.

To obtain a rough measure of the extent of this potential bias, I relied on FCC reports of the real average revenue per interstate conversation minute, population data, and real GDP. From this I simulated *terminating* interstate switched access minutes (that is, the number of minutes that the local carriers billed for completing ordinary interstate calls) for the period since 1992 based on movement in real prices, real GDP, and population.[10] The result is shown in figure 6-1. Note that the predicted and actual minutes track each other closely through 1996. Thereafter the series begin to diverge, with the predicted number of switched access minutes exceeding reported minutes by 2.5 to 6.9 percent in 1997–99, just before wireless competition for long-distance minutes begins in earnest.[11] This suggests that interstate minutes were being diverted in some manner after 1996, perhaps in the fashion alleged in the complaints lodged against WorldCom.

It is difficult to confirm this diversion hypothesis directly through an analysis of the margin of interstate rates over switched access charges. As table 6-2 shows, however, the margin fell from 8 cents a minute in 1992 to 6 cents a minute in 1995, but it was still 6 cents a minute in 2001 despite a continuing decline in long-distance carrier concentration.[12] Surely the added competition from wireless should have reduced these margins in 1999–2001. Thus it is likely that the FCC calculation of interstate revenue per minute is biased upward somewhat owing to an underreporting of switched access minutes. To adjust the FCC series on revenues per minute, I assume conservatively that the diversion began in 1997, which accounts for 2 percent of revenues, and that it grew by 2 percent a year through 1999 and then remained steady in 2000 and 2001. The resulting effect on predicted minutes can be shown as the difference between the predicted and "predicted-adjusted" minutes (figure 6-1).

Long-Distance Entry: Wireless and the Bell Companies

The 1996 act did little to change the long-distance market at first. It barred the Bell companies from interstate (more precisely, interLATA) long-distance services in their own regions until they satisfied a laborious state-by-state "checklist" designed to show that they were cooperating in allowing entry into their local markets.[13] As mentioned earlier, no Bell company would succeed in gaining approval to offer such service until the end of 1999. By the fall of 2002, however, the Bell companies had obtained approval in states accounting for more than 80 percent of their access lines, and by the end of 2003, they were allowed to offer service in all of the lower forty-eight states plus the District of Columbia.

Bell company entry was not the only major new source of competition in the U.S. long-distance market after 1996. An even more important event was the entry of wireless or "cellular" carriers into the market through nationwide plans. In May 1998, AT&T launched its "One-Rate" plan allowing subscribers to dial anywhere in the country from anywhere in the country at rates as low as 10 cents a minute, which was less than many residences were then paying for wireline long-distance service.[14] Nextel, Verizon (Bell Atlantic), and Sprint followed in the next year. These plans were extraordinarily successful, as reflected in accelerating wireless revenue growth and the slower growth of "roaming" revenues.[15] But their most important impacts were experienced by long-distance carrier rates.

Average interstate rates (excluding international calling) fell from $0.15 to $0.11 per conversation minute in 1992–98 and then declined to $0.07 and perhaps even less by 2002.[16] For most of the 1990s, wireless rates were far above wireline long-distance rates, but in 1998–99 wireless rates began to decline sharply (figure 6-2). By 2001 the average wireless rate was almost as low as the average long-distance carrier's rates, and many plans began to

Figure 6-2. *Interstate Long-Distance Rates and Wireless Rates, 1992–2002*

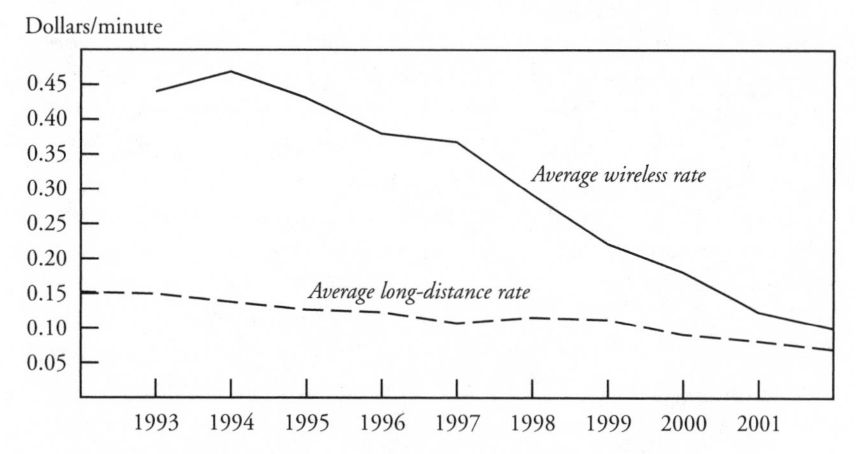

Source: FCC, *Telecommunications Industry Revenue Data, 2002* (March 2004); FCC, *8th Annual CMRS Competition Report* (July 2003).

offer large amounts—in some case unlimited amounts—of free wireless airtime after 9:00 P.M. on weekdays and any time on weekends.

Much of the early decline in long-distance rates was due to reductions in regulated access charges and universal service charges, not to competition. The average revenue per domestic interstate conversation minute fell from $0.149 in 1992 to $0.083 in 2001, a $0.066 reduction. During this period, switched access charges per conversation minute fell from $0.069 to $0.018, a $0.051 reduction. Thus access charge reductions contributed more than three-quarters of the reduction in interstate long-distance rates over this nine-year period, far more than may be attributed to other forces, including competition.[17] This is now changing due to wireless and Bell company competition.

The competition from wireless has not only reduced traditional long-distance carriers' rates, but it has begun to shift substantial traffic from these carriers to wireless operators. For several years, the decline in wireless rates and increasing subscriber penetration had been reducing the average cellular bill. However, the introduction of the nationwide pricing plans reversed this trend, presumably because increased usage offset the continuing fall in wireless rates.

For a rough estimate of the shift in long-distance minutes from wireline to wireless carriers, consult figure 6-1 once again. Given that the nationwide wireless plans had not become pervasive until mid- or late 1999, the gap between predicted and actual terminating access minutes (a good proxy for conversation minutes) until 1999 cannot be due to wireless competition.[18] However, the gap opened dramatically in 2000, rising to 30 percent of terminating wireline minutes. By 2002 actual minutes were 46 percent below the predicted values, suggesting that 46 percent of minutes had somehow disappeared from the traditional fixed-wire network.

Because the new wireless plans have diverted a substantial number of *originating* minutes from the wireline network as well, the analysis of the wireless carriers' effects on long-distance traffic should focus on *total* interstate switched access minutes, not just the terminating minutes shown in figure 6-1. When I simulate the movement of total interstate switched access minutes using real GDP, real interstate revenue per minute, and a time trend to capture the growth of "special access," I find that predicted switched access minutes exceed actual minutes by nearly 50 percent in 2002. The gap increases from just 45 billion minutes in 1999 to 465 billion minutes in 2002, an increase of 420 billion minutes. In this period of sharply declining interstate long-distance prices and rising GDP, total switched access minutes actually decline by 12 percent, from 553 billion minutes to just 486 billion minutes. Clearly, traffic is shifting to some other platform.

The effect of the introduction of the nationwide wireless plans in 1998–99 on wireless-wireline substitution is evident in the sharp rise in average wireless minutes beginning in 1998. Between 1994 and 1997, the average wireless subscriber's minutes of use per month varied narrowly between 117 and 125.[19] With the introduction of lower rates and nationwide calling plans, the average wireless usage rose to 185 minutes a month in 1999 and then to 427 minutes a month by 2002.[20] This suggests that all wireless calls, including long-distance calls, rose by more than 500 billion minutes a year.[21] The FCC reports that an estimated 26 percent of wireless minutes were attributable to interstate calls in 2002, or about 179 billion minutes in total. There are no estimates of the share of interstate minutes in wireless calling for 1999, but the FCC reports that in 2000 the share was just 16 percent.[22] In all likelihood, the share was substantially lower in 1999, when the national calling plans were beginning to spread. Assuming that 11 percent of wireless minutes were interstate in 1999, they would have totaled just 19 billion. Thus interstate wireless minutes rose by about 160 billion between 1999 and 2002, or about 40 percent of the 420 billion minutes that appear to have vanished from the wireline network. The

Table 6-3. *Average Household Bills from Long-Distance and Wireless Carriers*
Dollars/year

Year	*Long-distance bill*	*Wireless bill*
1995	250	82
1996	250	108
1997	305	129
1998	270	164
1999	257	205
2000	211	279
2001	176	351
2002	149	417
2003	122	492

Source: FCC, *Reference Book of Rates, Price Indices, and Household Expenditures for Telephone Service* (2004), table 2.6.

remainder must reflect the growth of e-mail, instant messaging, and perhaps even voice over Internet services.

The shift of long-distance service to wireless can be further demonstrated by the trend in household expenditures. Between 1995 and 1999, the average household's long-distance carrier bill fluctuated between $250 and $300 a year (see table 6-3). Beginning in 2000, however, the average residential long-distance bill began to plummet, falling by nearly one-half in just three years. At the same time, the average wireless bill rose at an average annual rate of 19 percent despite the sharp drop in wireless rates after 1998. As noted earlier, this increase in wireless spending reflected the rapid rise in wireless use, much of which must have been long-distance usage. Given the continuing sharp decline in long-distance rates, the slowing of the decline in access charges, and the substantial loss of minutes to wireless carriers, it can hardly be surprising that long-distance carrier profits and stock prices have plummeted.

The effect of wireless and Bell company competition on long-distance rates may be estimated through an analysis of monthly data from the consumer price index (CPI) for residential long-distance service. Consistent data are available from January 1998 through June 2004 and include observations that antedate the wireless national pricing plans and Bell company entry into long distance. A simple linear regression of this CPI index, deflated by the CPI for all items, on wireless minutes a month and the number of subscriber lines in states in which the Bell companies had received approval to offer interLATA service finds both variables statistically significant.[23] The results suggest that the rise in average cellular minutes a month from 136 in mid-1998 to 507 in mid-2004 caused about a 30 percent reduction in residential long-distance rates. Bell company entry reduced rates by an estimated 7 percent between March 2000 and March 2004.

Thus restricting Bell company entry into long-distance services as an "inducement" to open their networks cost consumers an estimated 7 percent more in long-distance rates in the first four years under the act and a declining percentage between 2000 and 2004 as the Bell companies were slowly being allowed to offer long-distance services from within their operating territories. Assuming that the total residential long-distance market in 1996–2000 was about $50 billion, this translates into a transfer from consumers to the failing long-distance carriers of approximately $3.5 billion a year and the loss of another $90 million a year in consumer surplus from lost long-distance minutes.[24] After 2000, this cost fell to perhaps $1 billion a year. Over the entire eight-year period, the total consumer cost of promoting local competition through blocking Bell company entry has been about $20 billion.[25] As noted in chapter 4, the benefits of this policy from enhanced local competition were substantially less.

Effect on Long-Distance Company Equities

The aggressive wireless-wireline competition developed just when the equity prices of long-distance companies began tumbling. An equity price index for the two largest long-distance carriers other than AT&T (see figure 6-3) shows a sharp rise between 1997 and 2000 and an equally sharp decline thereafter. (AT&T is excluded because it had become a large cable company with the acquisitions of TCI and MediaOne.) Surprisingly, the long-distance equities continued to rise after AT&T Wireless introduced its "One-Rate" plan in 1998. The wireless stocks also began rising briskly at this time as the heated competition and decline in roaming revenues caused by the nationwide plans reverberated through income statements.

In late 1999, two months before the first Bell company application for long-distance entry was approved under Section 271 of the new act, long-distance companies found their stock values slipping. Wireless equities did not begin their slide for another five months. Nor did they plunge into oblivion, as did many long-distance equities. Global Crossing and World-

Figure 6-3. *Wireless and Long-Distance Stock Prices, 1996–2003*

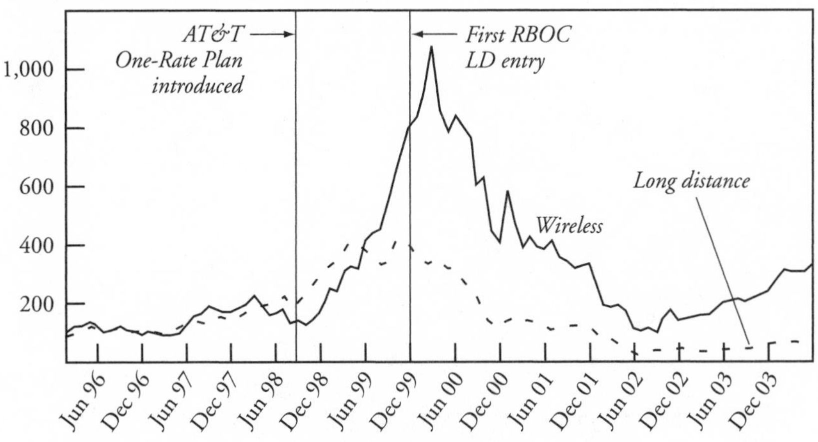

Source: Author's calculations and www.finance.yahoo.com.

Com, for example, entered bankruptcy. Global Crossing sold its Asian assets to other world telecom companies, and it and WorldCom were reorganized under U.S. bankruptcy laws. Sprint and Qwest still are solvent, in large part because of their local telephone operations.

At the end of 1999, AT&T, WorldCom (later MCI), Sprint, Qwest, and Global Crossing had a combined market capitalization of $472 billion and a combined market value of $568 billion. By the end of 2003, AT&T had sold or spun off its cable and wireless businesses, Qwest had all but abandoned the long-distance business, Global Crossing had been reorganized in bankruptcy, and WorldCom was preparing to emerge from bankruptcy. By mid-2004 three carriers—AT&T, Sprint, and the reorganized MCI—had a market capitalization of $41 billion and a market value of $87 billion. But even this latter total overstates the value of the long-distance business per se. The value of Sprint's long-distance assets is now essentially zero (see table 6-4 for details). Only AT&T retains much market value, but its equity share price has declined by more than 30 percent since it spun off its most valuable assets—namely, its cable operations. AT&T and MCI have a combined market value of about $32 billion in their long-distance operations despite having invested more than $40 billion in these operations in the past four years.

Table 6-4. *Estimated Market Value of the Leading Long-Distance Carriers, June 2004*
Billions of dollars

Carrier	*Market (market cap + book value of debt)*	*Non–long-distance assets*	*Long-distance assets*
AT&T	29.7	10.0 (local assets)[a]	19.7
MCI (WorldCom)	12.2	n.a.	12.2
Sprint	45.0	17.2 (local lines)[a]	0
		31.5 (wireless)[a]	

Source: www.quote.yahoo.com and company reports to the Securities and Exchange Commission.
a. Author's estimate.

Clearly, the financial markets place a very low value on these companies' long-distance assets, little more than the value of the cable television industry's home shopping networks.[26] There is even speculation that the sole remaining independent long-distance company of any value, AT&T, will be bought by one of the Bell companies or be sold to a group of private investors.[27]

Long-Distance Company Investment

It is possible that the long-distance companies could not have survived in a competitive environment under any circumstances once the Bell companies were free to offer in-region interLATA services. All the same, these long-distance carriers hastened their own demise by investing far too heavily in new facilities. As figure 6-4 shows, capital expenditures on long-distance networks quadrupled (in nominal dollars) between 1996 and 2000 despite falling capital equipment prices.[28] At the same time, their total revenues, net of resale, did not increase very much between 1996 and 1999 and then began to fall in 1999. As of October 2004, all the major long-distance carriers are reporting further sharp declines in long-distance revenues (see table 6-5).

Between 1996 and 2001, the five major long-distance carriers spent about $120 billion on their networks. By June 30, 2004, all but two of those networks were close to being worthless on the financial markets (table 6-4). There may have been a reason to increase capital spending somewhat between 1996 and 2000, given telecom's shift from voice to

Figure 6-4. *Long-Distance Carrier End-User Revenues and Capital Expenditures, 1996–2002*

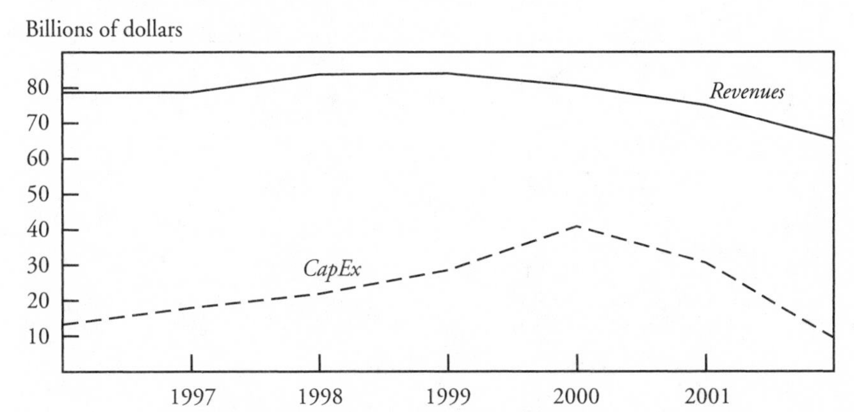

Source: FCC, *Telecommunications Industry Revenues, 2002* (March 2004); CreditSuisse/First Boston (2003).

data, but it is unlikely that any of the companies' investment plans were based on an assumption of *falling* revenues after 2000. Without substantial growth in overall demand, these companies could not survive the large traffic losses to the wireless carriers and the Bell companies alongside the sharp decline in average long-distance rates.

WorldCom's Reorganization: MCI Redux

The decline in carrier revenues shown in figure 6-4 has accelerated since 2001. The leading companies—AT&T, WorldCom (now MCI again), and Sprint—experienced a revenue fall of more than 11 percent a year between 2000 and 2004. The declines for AT&T and WorldCom (MCI) shown in table 6-5 are all the more surprising because they include the still rapidly growing local resale operations of each carrier using the UNE platform (see chapter 4). Given the continuing expansion of Bell company long-distance services as well as the continuing growth of wireless, there is no reason to expect revenues to stabilize. Indeed, the decline may even accelerate in the wake of the new bundled long-distance plans offered by the Bell companies and their expansion of activity in the large business or "enterprise" market from which they had been excluded by the interLATA restrictions. AT&T and WorldCom have dominated the enterprise market because the Bell

Table 6-5. Long-Distance and Related Revenues for Leading Carriers, 2000–04[a]
Billions of dollars/year

Company	*2000*	*2001*	*2002*	*2003*	*2004*[b]	*Average annual growth (percent)*
AT&T	46.8	42.2	37.8	35.6	31.3	–10.1
Sprint	10.5	9.9	8.9	8.0	7.4	–8.7
MCI (WorldCom)	39.1	37.6	32.2	27.3	22.9	–13.4

Source: Company reports to the SEC.
a. Includes some local service revenues, primarily through resale of ILEC services.
b. Estimates based on three quarters.

companies have been unable to offer services to these companies' diverse operations across LATA boundaries. This quarantine has now ended, and AT&T and WorldCom's share of this market is likely to decline substantially in the next few years.

Neither Sprint nor MCI (WorldCom) has been able to earn any measurable profits from its long-distance operations at recent revenue levels. Even after ridding itself of most of its debt through bankruptcy, MCI reported a $3.9 billion loss for the first three quarters of 2004.[29] Sprint reported a $1.7 billion operating loss for the first three quarters of 2004 in its long-distance operations.

Conclusion

The only reasonable conclusion that can be drawn from the empirical evidence on the U.S. long-distance sector is that independent long-distance companies are disappearing. Given the onslaught of competition from wireless companies, the recently emancipated Bell companies, and—now—VoIP, the revenues of the long-distance carriers are falling at the rate of 10 percent a year or more. Unless these companies can develop a full array of telecom services, including local access, broadband, and wireless, they are unlikely to continue to operate profitably. Without their own local networks, they will be unable develop these services, and it is surely too late for them to enter (or reenter) the wireless business. AT&T has a substantial investment in fiber rings in large cities to serve the largest business customers, as does MCI, but neither can reach the mass market with its own facilities. Unfortunately, none of the long-distance companies is investing any further in its own local infrastructure. Thus they are in the process of allowing most of their business to migrate to wireless carriers, incumbent local exchange carriers, and—eventually—cable television companies.

For much of the past eight years, the long-distance carriers were protected from Bell company competition in interLATA markets by the 1996 Telecommunications Act. Keeping the Bell companies out of the long-distance market cost consumers $20 billion. By the end of 2003, this protection had vanished, and the burden on consumers had been lifted.

Much has been made of the accounting scandals that have gripped this sector, such as those that led to the collapse of Global Crossing and WorldCom and converted Qwest back to a local exchange company. These scandals undoubtedly exacerbated the sector's problems and may even have contributed to the loss of switched access revenues for the incumbent telephone companies. Even so, excessive investment at the dawn of the new competitive age was the most important cause of the long-distance debacle of 2001–04. Now, with sharply declining revenues, the independent long-distance companies cannot remain viable competitors in the telecommunications marketplace in the long run, even after correcting their accounting practices in the past decade. The scandals merely accelerated their demise.

7 The Rapid Growth of Wireless Telecommunications

The number of wireless or "cellular" subscribers in the United States will soon exceed the number of traditional telephone lines.[1] Most subscribers now have a choice of five national cellular carriers. Nevertheless, regulators continue to view the provision of "local" telephone access and exchange services as a local monopoly, thereby requiring government regulation. But if nearly every current user of ordinary local telephone service will soon be carrying a wireless handset, and if this wireless service can be purchased from five national vendors, how can local telephone companies enjoy any market power in traditional voice services?

Recent Trends in Wireless and Local Wireline Access

American consumers clearly have a variety of options for sending and receiving voice and data messages over the global telephone network. For example, they can subscribe to the local incumbent telephone company's service, and in most larger communities they can subscribe to a new competitive local exchange carrier (CLEC) or purchase telephone services from the local cable television company. Then there are the wireless services such as Sprint PCS, Cingular, Verizon, T-Mobile (VoiceStream), or Nextel, which offer further choice.[2] Pay telephone services are also available, although in general they cannot compete with a home or wireless connection.[3] All of these services allow the consumer to originate and receive local, intrastate, interstate, and international calls.

The first analog cellular service began in 1983, but digital wireless services did not become widely available until 1995. Since that time, wireless service has grown tremendously, while the price of using it has plunged. As figure 3-8 showed, the gap between fixed-wire access lines and wireless subscribers is narrowing rapidly. The biggest surprise in figure 3-8 was not the steep continuing rise of wireless subscriptions, but the sudden decrease in traditional telephone access lines beginning in 2001. For the first time in modern history, the number of traditional lines declined.

The incumbent telephone companies' end-user access lines shrank from 177.6 million at the end of 2000 to 151.8 million at the end of 2003, a decrease of approximately 26 million lines. By contrast, the new competitive carriers (CLECs) reported an increase of less than 15 million lines in the same period.[4] Nearly 80 percent of the incumbents' decline was in residential lines and small business lines, and a similar percentage of the CLEC increase was also concentrated in this residential and small business category.[5] One might attribute the sudden decline in wire-based lines to the 2000–01 economic slowdown, except that telephone subscriptions did not plummet during the last U.S. recession, in 1990–91. Nor did lines decline in earlier post–World War II recessions. One of the new culprits is the decline in second lines due to the growth in broadband. Another is likely the development of wireless as an alternative to conventional telephony.

The Development of Wireless Services

Before 1983 few people considered acquiring a mobile telephone because of its high cost and the low quality of service arising from its inefficient use of the electromagnetic spectrum. However, a new cellular technology in the 1970s allowed much more efficient use of the spectrum because the same frequencies could be used simultaneously in different "cells" within a given metropolitan area. This new technology led the Federal Communications Commission (FCC) to allocate two 20-megahertz bands (later increased to 25-megahertz bands) in each local area, one to the incumbent local company and one to a second, independent carrier. It would take almost ten years before licensing policies were finalized and cellular service could begin. In 1983 the first cellular service began operation in Chicago.

The antitrust decree that divested the Bell operating companies from AT&T in 1984 conveyed the local cellular licenses in their franchise areas to the divested Bell companies.[6] These companies vied with the second, independent carrier in local duopolistic markets that were free from any entry threat for the next decade. In 1993, however, Congress included in the Omnibus Budget Reconciliation Act an instruction to the FCC to begin auctioning spectrum for new wireless services, thereby creating a major increase in competition.[7] Wisely, it also instructed the FCC and the states to forbear from regulating wireless rates unless the carrier enjoyed "market dominance." The result is that a decade later wireless rates are now essentially unregulated and very low.[8]

In 1994 the commission launched the process of auctioning spectrum by offering two 30-megahertz bands for the new personal communications services (PCS), which would evolve into digital cellular services. Subsequently, the FCC auctioned off another 60 megahertz for this purpose, and an entrepreneurial carrier, Nextel, succeeded in persuading the commission to allow it to convert 10–15 megahertz of spectrum from another wireless service to its own digital cellular service.

At first, the FCC viewed wireless as a strictly local service and issued licenses for narrow metropolitan areas. Over time, the local licensees combined into larger regional and, eventually, national carriers through an undoubtedly costly process that probably slowed the growth of U.S. wireless. McCaw began this trend with a large number of acquisitions in the 1980s and eventually sold its entire cellular business to AT&T, which had been left without wireless operations by the 1984 divestiture. Subsequently, Vodafone, Europe's largest wireless carrier, purchased Air Touch, which had earlier combined the cellular operations of Pacific Telesis and U S West. Vodafone and Verizon then formed a joint venture to create a national wireless carrier. Sprint purchased a large number of licenses in the auctions and created a national carrier, Sprint PCS. SBC and Bell South combined their cellular operations to create a national carrier, Cingular. After several prior mergers created Voice Stream, it was acquired by Deutsche Telecom and is now marketed as T-Mobile. Nextel gradually expanded its network across the country by acquiring spectrum in noncellular spectrum bands to create a sixth national carrier. The number of national carriers was reduced to five by Cingular's purchase of AT&T Wireless in 2004 at a cost of $41 billion plus the assumption of AT&T Wireless's debt, and the number may decline further to four if Sprint's merger with Nextel is completed in 2005.[9]

This evolution of wireless services occurred in an environment of little direct government regulation of rates or technology. The FCC controls the auctioning of spectrum and requires wireless carriers to interconnect with other carriers at rates based on the reciprocal compensation rates negotiated among local carriers in each state. Otherwise, wireless rates are free of regulation because the 1993 legislation authorizing auctions forbids it unless a carrier has market dominance. Nor has the U.S. government attempted to standardize wireless technology as the European Union has done.[10]

As commercial wireless services developed, carriers were forced to deal with subscribers who "roamed" outside their own franchise areas. At first, each market had a limited number of wireless carriers, and they charged each other very high rates for serving one another's roaming customers. These rates were in turn reflected in retail roaming rates. Between 1995 and 1998, roaming revenues accounted for 10–15 percent of all wireless revenues.[11] The high roaming rates pushed carriers to expand their national networks, and increased competition among the national carriers began to place downward pressure on roaming rates and revenues. In 1998 the share of roaming revenues in total revenues began to decline, and by 2003 it had fallen to about 4 percent of revenues.[12]

In 1998 AT&T also introduced its national "One-Rate" plan, allowing subscribers to call from anywhere to anywhere in the United States at the same rate. Over the next year, the other carriers followed suit, and today each carrier provides an increasing share of its service through such plans.

The Growth of Cellular Usage

Wireless service is no longer simply a mobile version of local voice telephone services. When AT&T announced its One-Rate national calling plan in 1998, it sparked a revolution in U.S. voice communications. In its first fifteen years of existence, the cellular telephone sector encountered declining revenues per subscriber because of declining rates and the enrollment of incremental subscribers who use a cell phone less intensively than the early adopters.[13] As mentioned in chapter 6, between 1994 and 1997 wireless usage averaged about 120 minutes a month, but by 2002 it had more than trebled to 427 minutes a month.[14] This surge reflects a steady shift in long-distance calling from the traditional wire-based network to wireless and, to a much smaller extent, the growth of new wireless applications, such as text messaging, e-mail, and direct Internet connections.

Today each of the national carriers offers a variety of calling plans and service feature packages extending beyond the voice service just described in chapter 6. The current 2-G technology can provide a range of services from paging, text messaging, personalized graphics, and ring tones to Internet browsing. Subscribers may pay as little as $15 a month for such features, which is what Sprint PCS charges for a variety of messaging, video mail, and web services.[15] In addition, a high-speed third generation (3-G) of services is slowly making its way into the United States.[16] These not only enhance the value of a subscription but also add to the average minutes of use.

Because traditional cellular services can now be used to connect to the Internet, the local wire-based telephone line is no longer necessary even for Internet services. A household can obtain Internet access at speeds comparable to those provided by dial-up connections by connecting its computer, generally a laptop, to a wireless service. Alternatively, one can browse the Internet directly through the handset or other portable device, such as a personal digital assistant (PDA). By the end of 2002, an estimated 12 million wireless subscribers were using some form of wireless data service.[17] In addition, consumers in many major metropolitan areas are now able to access Verizon Wireless's new broadband service for about $80 per month.

Competition in the Wireless Sector

Because regulators viewed cellular service as a high-value complement to existing local services, they approached it from a very narrow geographic perspective, initially awarding licenses to two carriers in each local market in the country. When the FCC began to auction new spectrum for cellular services in December 1994, it continued in the same manner, auctioning two 30-megahertz bands in each of fifty-one major trading areas (MTAs). In its next two major cellular auctions, involving a total of 60 megahertz, it divided the country into 493 basic trading areas (BTAs), roughly corresponding to census metropolitan statistical areas.[18] These decisions were taken before it became obvious that cellular service would evolve into one that required a national footprint because many users travel between the MTAs or BTAs and, particularly after 1998, use their cell telephones for a large share of their long-distance calling.

The FCC's auction process raises a number of well-publicized problems, particularly with respect to bidder preferences for small businesses.[19] Yet

Table 7-1. *U.S. Wireless Subscribers, December 31, 2003*
Millions

Carrier	*Subscribers*	*Share (percent)*
AT&T Wireless[a]	22.0	13.8
Cingular	24.1	15.2
Nextel	14.0	8.8
Sprint PCS	15.9	10.0
T-Mobile	14.0	8.8
Qwest	0.9	0.6
Verizon	37.5	23.6
Others	30.3	19.1
U.S. total	158.7	100.0

Source: Company websites; Cellular Telecommunications Inudstry Association, *Semiannual Wireless Survey* (December 2003).

a. Acquired by Cingular in October 2004.

the decision to carve the country into so many areas for spectrum auctions has received little attention, even though it created a complex of roaming agreements among the hundreds of licensees across the country. Each carrier had to establish agreements with carriers in every other area of the country to allow its subscribers to travel out of their home territory and still be able to use their cell phones. Since carrier technologies vary greatly, there may have been only one or two carriers in each market to negotiate with, or subscribers may have been forced to purchase "dual-mode" or "tri-mode" handsets to connect with carriers using different frequencies or technologies. This vast array of expensive roaming arrangements arose because each licensee was given market power in its own area.

Wireless service evolved from a largely local service to a national service as each major carrier developed its own national network to avoid excessive roaming charges. Carriers began to merge, acquire licenses from other carriers, and seek additional licenses in various FCC auctions. By the end of 2003, the five largest national carriers had 80 percent of all cellular subscribers (see table 7-1). Moreover, many of the smaller companies are affiliated with these national carriers, and even Qwest has begun to offer its service over Sprint's national network.[20] Thus the wireless sector in the United States has been transformed from a large number of small, local operations into five large, national carriers who compete actively among themselves and with traditional wire-based carriers for customers.[21]

The Relative Prices of Wireless and Wireline Service

In the United States, local residential service is offered at a regulated flat rate that does not vary with the number of calls or minutes of use in a local calling area. Wireless services do not offer free calling except in off-peak hours. Wireless connections have become enormously popular in part because of the substantial real decline in average wireless rates, shown in figure 7-1. Between 1996 and 2002, real cellular prices per minute declined at an annual rate of 23 percent, and these prices now include local and national calling because all national wireless carriers currently offer a single plan for local and national calls.[22]

By contrast, real revenue per minute for interstate long-distance services that are dialed over traditional wire-based networks fell by only 9.5 percent a year between 1996 and 2002, despite a 25 percent annual decline in real switched access charges.[23] The real consumer price index (CPI) for residential landline long-distance calls, including intrastate long distance, fell by only 7.2 percent a year between December 1997 and December 2003.[24] Finally, the cost of subscribing to a local line has actually increased since 1996, largely because of a regulatory shift in the burden of non-traffic-sensitive costs from per minute long-distance access charges to fixed per line charges. In real terms, this increase has amounted to about 0.3 percent a year.[25]

Clearly, the cost of using a wireless telephone has declined rapidly in comparison with the cost of leasing a local landline and using it. Local calls are still free to most local residential wireline telephone subscribers, but the cost of subscribing to the line is rising slowly. By contrast, the real cost of sending (and receiving) local and long-distance calls over a wireless handset has been falling at double-digit rates. Surely this suggests at least the possibility that sufficient consumers view a wireless service as a substitute for wireline service to make it difficult for traditional telephone companies to exercise market power.

Evidence of Wireless-Wireline Substitution

Ideally, one would use econometric analysis of individual subscription decisions to estimate the degree of current wireless-wireline substitution. The determinants of such decisions would be the relative prices of wireline and wireless access, the price of using the network through the two technolo-

Figure 7-1. *Real Long-Distance and Cellular Rates, 1993–2002*

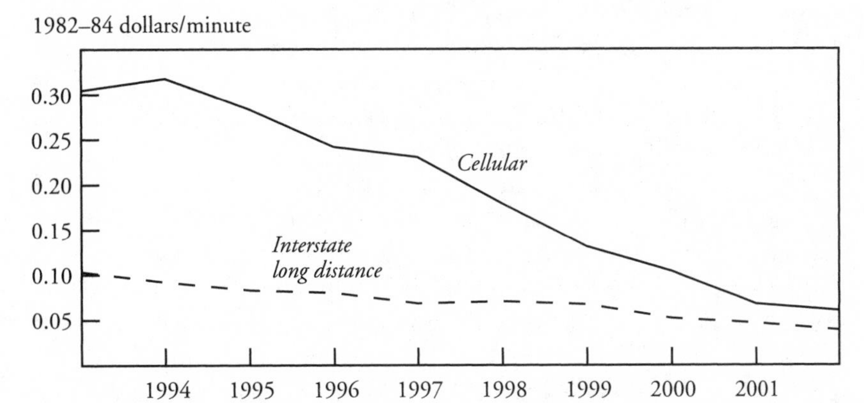

Source: For long distance: FCC, *Telecommunications Industry Revenues, 2002* (March 2004); for cellular: FCC, *Eighth Annual CMRS Competition Report,* appendix D, table 9; for deflator: BLS, *Consumer Price Index.*

gies, household income, and various demographic variables, particularly the age of the consumer. It is well understood that teenagers and young adults are much more likely to use cell phones than are older consumers with similar incomes. Unfortunately, individual subscriber data are not easily obtained; most disaggregate data on telephone usage is obtained at the household level, and such data are generally not available for households without a wireline connection.

In a recent study of wireless-wireline substitution using U.S. household data, Mark Rodini and his colleagues found the cross-price elasticity between wireless subscription and the price of a second fixed-wire line to be 0.18 and 0.13 in 2000 and 2001, respectively.[26] They also calculated the own-price elasticity of wireless subscription to be –0.43. These results are for at least three years ago, when the number of wireless subscriptions was much lower and the wireless pricing plans were likely much less generous than those available in 2004. Therefore the relatively low estimate of cross-price elasticity should probably be viewed as a lower-bound estimate of today's potential wireless-wireline substitution.

Another problem in estimating the effect of relative prices on the choice between wireless and wireline access is that even though wireless rates vary enormously because of the myriad plans offered by the carriers, these plans are essentially the same throughout the country. Therefore virtually every consumer in the United States faces the same prices for wireless service. Fortunately for this exercise—although not necessarily for consumers—local rates vary substantially across states and even within states. This variance allows us to test for the effect of prices on the choice between a wireless and a traditional wireline telephone.

Using FCC semiannual data on wireless subscriptions across states, I tried to determine if wireless subscriptions are plausibly influenced by the level of wireline rates through a pooled time-series, cross-section regression of wireless penetration rates in December 2000, 2001, 2002, and 2003.[27] As independent variables I used the mean income per capita, the share of the population living in urban areas, and two fixed-wire price variables: the price of basic local residential service and average intrastate long-distance rates.[28] The coefficient for the fixed-wire local price variable was generally significant when no year-specific dummy variables were included.[29] However, when year-specific dummy variables are included in the regression, the coefficients on both price variables become statistically insignificant. Given the imprecision of the price data and the aggregated data being used, I simply cannot find a statistically significant cross-price elasticity of demand. Evidence on substitutability of wireless for fixed-wire service may be available elsewhere, however.

For instance, it may be that individual households are disconnecting from wireline service and relying solely on wireless services. Such a pattern has already been established for Scandinavian countries, where wireless penetration is much greater than in the United States. Perhaps the recent decline in total U.S. access lines indicates the beginning of such a phenomenon in the United States. If so, is the competition from wireless now sufficient to permit deregulation of local access service?

The only direct public evidence on recent subscription decisions by U.S. households comes from the Census Bureau's monthly *Current Population Survey* (CPS). Three times a year, in March, July, and November, the bureau asks each of 56,000 households if it currently subscribes to or has access to local telephone service. Over the past decade, the share of U.S. households with a telephone has remained remarkably stable at about 94 percent.[30] Even in a year of mild recession, 2001, the share did not decline. Could these two divergent trends be explained by the substitution of wireless for wireline access? The evidence has been unclear because the census questionnaire began to ask each respondent if it had only wireless or only wireline service only in November 2001. From the first survey's fragmentary results, it appears that only 1.2 percent of responding households had only wireless service. Because of design problems, the question was dropped from the CPS for a number of years. In February 2004, however, the question was included once again, and the estimated share of households with only wireless service has now risen to 6 percent.[31]

The CPS data confirm the results from private market research by companies such as J. D. Power and Associates, which collects data on household subscriptions to cellular service. In its 2003 survey, J. D. Power found that approximately 3 percent of U.S. households have a cellular telephone but no wireline service.[32] In a recent survey of household *preferences,* Primetrica and Ernst and Young found that between 42 and 58 percent of U.S. households would be willing to buy a wireless plan that offers 600 "anytime" minutes per month for $50 and drop their fixed-line service. Between 29 and 44 percent of the households surveyed would be willing to pay $60 for 2,000 minutes a month and drop their fixed-line service.[33]

Given the enormous shift to wireless in long-distance calling (see chapter 6), it is not surprising that some households are now dropping wireline service altogether, and many more say that they would be willing to do so. Unless the household relies on a digital subscriber line (DSL) for broadband service, there is little need to keep a wireline connection once household members have a wireless phone.[34] As service quality on wireless connections improves, wireless-wireline substitution will become even more intense.

A Competitive Equilibrium?

None of the five remaining national carriers shown in table 7-1 has a dominant share of wireless subscribers. Even after adding Qwest's subscribers to Sprint's total, the Herfindahl-Hirschman index (HHI) of concentration for the five largest carriers is only about 2550.[35] Given the fact that wireless services compete with wire-based services, this index overstates any market power of the carriers. Moreover, given the dynamic expansion of the market, the rapid technological change in the sector, and the expanding array of services, it is unlikely that a market structured with approximately four equal-size competitors would generate collusive behavior.

Concerned that the U.S. wireless sector is *too* competitive, many financial analysts have long advocated mergers among some of the participants. Because wireless services are growing rapidly, carriers have been forced to expand their network capacity and spend large sums on attracting and keeping subscribers. As a result, many wireless carriers continue to generate rather modest profits, and some even report losses in their current income statements. The sharp decline in wireless stock prices in 2000–02 convinced many observers that the industry appeared unable to support six national players before Cingular's purchase of AT&T Wireless. Since that acquisition, wireless equity prices have rebounded substantially.

I do not attempt to model the national delivery of wireless service in the United States to determine whether a market equilibrium with five carriers is sustainable. Rather, I examine the recent values of wireless service companies to determine whether they are above or below the reproduction costs of their assets, that is, whether their "Tobin q" values differ from 1.[36] The cost of building a wireless services company includes the cost of acquiring the spectrum, the cost of building network facilities, and the cost of acquiring customers.

Spectrum

Since 1993 wireless providers have had to purchase spectrum through the FCC's wireless auctions or by buying licenses in the secondary market. The cost of such spectrum is typically reckoned in terms of megahertz POPs, equal to one megahertz of spectrum that can reach one person in the country. Given that the United States now has a population of 292 million and that cellular service typically uses 30 megahertz, a fully national service (including Alaska and Hawaii) would require 8,760 million megahertz POPs. The price per megahertz POP for each successive FCC auction is shown in table 7-2.[37] These prices first surged in the C auction, which allowed eligible "small entities" to defer payments over time. Many of these winning bidders subsequently defaulted. The peak prices were realized in the C and F re-auction, the results of which were subsequently nullified by a court ruling that the major defaulting original winning bidder, NextWave, could not be stripped of its spectrum. The exuberant financial markets propelled the value of spectrum to $4.19 per megahertz POP in that auction, but stock and spectrum prices have receded substantially since then.

In recent activity, NextWave has been selling much of its spectrum to other wireless operators. The price of a very large share of this spectrum, which was sold to Cingular in August 2003, was approximately $1.60 per megahertz POP, or very close to the price of $1.33 per megahertz POP

Table 7-2. *Average Price per Megahertz POP for FCC Broadband PCS Auctions*

Auction number	*Auction name*	*Date started*	*Net revenues (billions of dollars)*	*Net revenues per Megahertz POP*
4	AB	December 1994	7.034	0.52
5	C	December 1995	9.270	1.33
10	C Reauction	July 1996	0.904	1.94
11	D, E, F	August 1996	2.523	0.33
22	C, D, E, F Reauction	March 1999	0.412	0.10
35	C, F Reauction	December 2000	16.857	4.19

Source*:* Auction data for auctions 4, 5, 10, 11, and 22 from FCC, *Fifth Annual Report on the State of Wireless Competition,* table 1B; data for auction 35 from Federal Communications Commission Public Notice, *C and F Block Broadband PCS Auction Closes; Winning Bidders Announced; Down Payments Due February 12, 2001, FCC Forms 601 and 602 Due February 12, 2001; Ten-Day Petition to Deny Period,* DA.

realized during the original C auction.[38] I therefore use $1.50 per megahertz POP as the long-term "equilibrium" price of spectrum.

With current technology, a cellular licensee uses 30 megahertz POPs of spectrum for each potential person reached by its signal. Assuming that the industry can enroll a present value of 60 percent of the people it reaches and that these subscribers are shared by five national carriers, these 30 megahertz POPs of spectrum will deliver an expected 0.12 subscriber. Thus the licensee must purchase 250 megahertz POPs per expected subscriber. At a price of $1.50 per megahertz POP, the cost of spectrum is $375 per subscriber.

Network Expenditures

Between 1985 and the end of 2002, wireless service providers spent a total of $126.9 billion on capital facilities to deliver cellular services.[39] This is a simple sum unadjusted for inflation or depreciation. Some early facilities have obviously been replaced by now. In 2001–03 carriers spent $56.2 billion on capital facilities and attracted 49.2 million additional subscribers.[40] Thus they spent about $1,150 per additional subscriber enrolled, but a substantial amount of this expenditure was likely directed to upgrading their network and replacing old assets. At most, the cost of building a (2G) network capacity at today's standards is likely $800 per subscriber.

Table 7-3. *Market Values of Independent Wireless Carriers, December 2003*

	AT&T Wireless	*Nextel*	*Sprint PCS*[a]
Subscribers (millions)	22.0	14.0	15.9
Market capitalization (billions of dollars)	33.2	29.6	6.0
Long-term debt (book value in billions of dollars)	10.5	10.2	16.7
Total market value (billions of dollars)	43.7	39.8	22.7
Market value per subscriber (dollars)	1,986	2,843	1, 428

Source: Company reports; www.finance.yahoo.com.
a. June 2003 estimate.

Customer Acquisition Costs

Each carrier reports its current cost of acquiring customers through its marketing operations. In the second quarter of 2003, these costs averaged approximately $350 per new subscriber.[41] In addition, the carriers must invest in marketing expenditures to replace the subscribers lost to "churn," that is, the customers disconnecting from their service. Since carriers currently experience a churn rate of about 30 percent a year, the present value of the cost of churn over the next five years, at a 20 percent cost of capital, is another 0.95 times $350, or $330. The total cost of acquiring and keeping a subscriber is, therefore, about $680.

Market Values versus Total Costs per Subscriber

At the end of 2003, the two largest independent wireless carriers, AT&T Wireless and Nextel, had a market value—equity plus (book value of) long-term debt—of $2,000 and $2,800 per subscriber, respectively.[42] Six months earlier, before Sprint eliminated its Sprint PCS tracking stock, Sprint PCS had a value of $1,400 per subscriber (see table 7-3). My estimate of the reproduction cost of the assets to serve that customer is $375 for spectrum, $800 for capital facilities, and $680 for customer acquisition, for a total of $1,855.

This estimate is slightly less than the market value per subscriber of AT&T Wireless and substantially lower than the value per subscriber for Nextel, but it is greater than the value of Sprint PCS. Nextel is an innovative carrier whose market value may reflect the financial market's belief that it has substantially more growth potential than the other carriers, and

Sprint's lower value is likely due to its problems in obtaining as good geographical coverage as that of its major competitors.[43]

Thus it appears that at spectrum values of $1.50 per megahertz POP, the value of U.S. wireless carriers is fairly close to the cost of building a system and attracting subscribers. This should not be surprising, for the market signals that drive the recent spectrum market transactions should be consistent with those affecting stock market values.

A Comparison with Other Countries

It is frequently alleged that the United States lags behind other countries in developing an enlightened wireless policy because it has much lower penetration (subscribers per capita) than many countries in Europe and Asia (see table 7-4). This low U.S. subscriber penetration is generally attributed to two factors: (1) the lack of the "calling party pays" regime employed in most of the world, through which the caller pays for the entire call; and (2) the failure to adopt a single technical standard. However, it is now evident that the U.S. (and similar Canadian) policies in these areas are superior to those of most other countries.

First, by requiring *both* parties to share in the cost of a call, the U.S. system allows competition to drive the prices for both originating and terminating the calls on mobile handsets. In Europe, in particular, regulators have been struggling with the regulation of wireless termination charges that they claim are subject to "monopoly" power. This alleged power derives from the fact that subscribers have no incentive to discipline carriers who attempt to charge more than a competitive market rate for terminating the call because the subscribers do not pay for receiving calls. In the United States, a party generally pays the same amount for receiving and for originating calls. Competition keeps these rates low, as the last column in table 7-4 demonstrates.[44]

Second, "calling party pays" is not likely the most important explanation for greater mobile wireless telephone penetration in Europe or Asia. These countries typically have very high local calling rates on their traditional wireline telephone networks, particularly during peak hours.[45] By contrast, the United States and Canada have large, free local calling areas. As a result, the incentive to shift from the landline phone to a wireless phone is much weaker in these two countries.

Third, the claim that a common standard across countries in Europe increases the attractiveness of wireless is surely overstated. While it is true that

Table 7-4. *Mobile Wireless Performance Indicators, 2003*

Country	*Penetration (subscribers/ 100 persons)*	*Who pays?*	*Minutes of use per month*	*Average revenue per minute (dollars)*
United States	54	Both parties	557	0.10
Canada	41	Both parties	296	0.12
Finland	92	Caller	243	0.18
France	68	Caller	174	0.23
Germany	79	Caller	75	0.33
Italy	99	Caller	116	0.25
United Kingdom	91	Caller	147	0.22
Japan	67	Caller	161	0.31
South Korea	70	Caller	311	0.10
Australia	78	Caller	176	0.20

Source: FCC, *9th Annual CMRS Competition Report* (September 2004), appendix A, table 11.

U.S. carriers use three or four different standards, each uses the same standard across all fifty states and offers a roaming-free national calling plan. One can take a GSM telephone across borders in Europe, but only at severe financial risk because intercountry roaming rates are extraordinarily high. Frequent travelers may have accounts in each country and a separate chip (or "SIM card") for each country, but such a practice is surely much less convenient than the situation in the United States.

Finally, one cannot ignore the striking result that emerges from table 7-4. Wireless rates are lowest and usage is greatest in the United States and Canada.[46] In the United States, in particular, the low, national rates have propelled cell phone usage to extremely high levels by international standards. If European policies are better, why are their prices much higher and their usage much lower than those found in the United States?

Conclusion

The U.S. wireless sector is growing rapidly, both in terms of subscribers and in minutes of use. A highly competitive set of five carriers introduced national pricing plans that have diverted a very large share of minutes from the fixed-line network. This competition developed in an environment of very little regulation. As a result, prices reflect network costs, not the regulators' desire to cross-subsidize various services. Recently, many wireless subscribers have decided to drop their fixed-line service altogether. Although there is no firm econometric evidence on the degree of wireless substitution for fixed-wire service, wireless carriers have captured a very large share of the long-distance market in recent years. Soon this wireless-wireline competition is likely to lead to a demand for deregulation of rates on the fixed network as well.

8 The Broadband Revolution

Although the Internet did not develop commercially until 1990 and the World Wide Web was not available until 1991, it did not take long for these forces and household ownership of personal computers to create a "Nation Online."[1] Just twenty years ago, only 8 percent of U.S. households had a computer; but by 2003, 61.8 percent had one.[2] Residential Internet connections lagged behind the growth in household computers at first, but by 2003 more than 50 percent of households were connected to the Internet (figures 8-1 and 8-2). According to the Census Bureau, between 1998 and 2003 the share of U.S. households connected to the Internet nearly doubled to 54.6 percent, driven by e-mail, e-commerce, and the explosion in the availability of easily accessible information on the Internet. Unfortunately, more recent census data on household Internet use are sparse, but a private research company estimates that there were 79.2 million residential U.S. subscribers in the first quarter of 2004, equal to about 72 percent of U.S. households.[3]

This rapid diffusion of a new communications medium led to a sharp rise in websites offering a variety of content, including recorded music, film clips, and electronic games. In a very short period, however, Internet users became frustrated by the "worldwide wait" that occurred when they connected to the Internet through ordinary modems at speeds of 50 kilobits per second (kbs). As a result, they began to seek higher-speed connections once the price of the electronic equipment required to deliver such services fell to affordable levels.[4] Larger businesses had been able to obtain access through

Figure 8-1. *U.S. Households with a Computer by United States, Rural, Urban, and Central Cities, Selected Years*

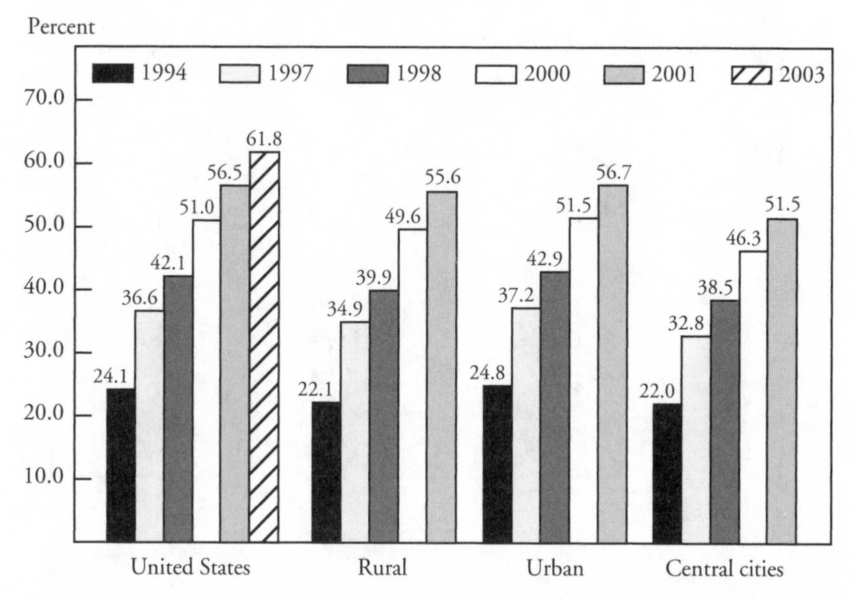

Source: National Telecommunications and Information Administration (NTIA) and Economics and Statistics Administration (ESA), U.S. Department of Commerce, using U.S. Bureau of the Census Population Survey supplements.

fast T-1 lines (1.544 megabits per second, or Mbs) for some time, but it was not until 1998 that higher-speed connections became available to households and small businesses through digital subscriber lines (DSL) over ordinary copper telephone loops and through cable modems provided by cable television companies.

Alternative Broadband Technologies

Residences and small businesses can obtain high-speed Internet access not only through DSL and cable modems but also through wireless and satellite devices.[5]

Digital Subscriber Line

Copper wires ("loops") that extend from a telephone company's central office or a remote terminal now serve virtually every household and business

Figure 8-2. *Percent of U.S. Households with Internet Access, by U.S., Rural, Urban, and Central Cities, Selected Years*

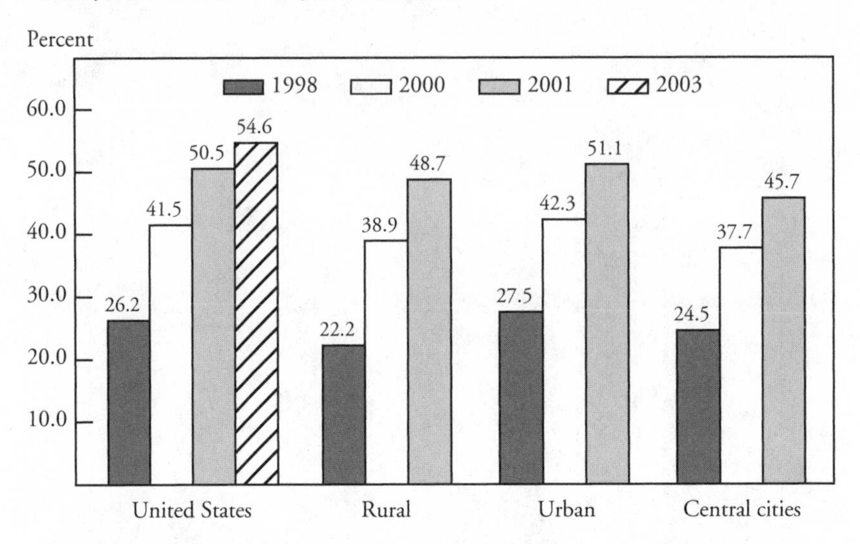

Source: NTIA and ESA, U.S. Department of Commerce, using U.S. Bureau of the Census Population Survey supplements.

in the United States. Using equipment much like an ordinary modem, telephone companies can transmit data at a high speed over these copper wires. DSL modems can transmit data at far higher rates than is possible over voice-grade connections because they can take advantage of capacity in the copper wire that is not used for voice communications. Asymmetric digital subscriber line (ADSL) service refers to a DSL system in which transmission capacities differ in the two directions of transit—with downloads arriving at much greater speeds than upstream communications. ADSL is particularly useful for web browsing or distributing audio or video programming, and it can operate on the same loop used for ordinary telephone calls. Thus by installing extra equipment at each end of an existing subscriber loop, or telephone line, a telephone company can deliver a high-speed access service on the same line that delivers standard telephony.

Unfortunately, DSL is severely limited by the quality and length of the copper loop. DSL connections can be restricted or even made unworkable if the copper loop has been modified with loading coils, which improve voice transmission on longer loops. More important, the ability of copper telephone loops to carry high-speed data signals declines with distance. ADSL permits downstream speeds of up to 8 megabits a second and upstream speeds of up to 640 kilobits a second, but speeds are much lower over the average loop. In fact, most U.S. ADSL subscribers realize downstream speeds of less than 1 megabit a second. ADSL modems do not operate reliably over copper loops longer than approximately three miles, a significant constraint in a country with a dispersed population.

In order to offer DSL service, a telephone company must install a DSL modem at the customer premises and at the central office or the remote terminal. Usually the modems in central offices come in assemblies of multiple modems called digital subscriber line access multiplexers (DSLAMs). The DSLAM must then have a high-speed connection to the rest of the Internet.

To improve the quality of all telephone services and to be able to offer DSL services in areas of low to moderate population density, telephone companies have extended fiber optics to remote terminals, where the light waves are converted to electrical signals and transmitted to the subscriber over much shorter copper lines. This architecture is known as digital loop carrier (DLC) and requires substantial investment in fiber optics and electronics at the remote terminals. In recent years, the incumbent telephone companies have been deploying this technology so as to offer DSL to more subscribers, but this deployment has arguably been delayed somewhat by uncertainty over whether the incumbent companies must share these facilities with new entrants.[6]

Cable Modems

Cable television systems use coaxial cable to distribute television signals from their central facilities, or "headends," to subscriber premises. Historically, these were one-way distribution systems built with coaxial cable and microwave radio systems, but in the 1980s they evolved into hybrid fiber-coaxial cable or "HFC" systems, using fiber optics to the subscriber's neighborhood and coaxial cable the rest of the way. The HFC cable system design improved the quality and reliability of cable systems and allowed the systems to increase their total bandwidth, some of which is now used for broadband Internet services. These HFC designs reduce the amount of coaxial cable in the upstream path that signals follow from the home to the network. Because the coaxial part of the network limits the capacity of the return channel, shortening the coaxial cable in the return path increases the capacity and reliability of the upstream path.

The cable industry and its suppliers initially experimented with many forms of two-way data communications over cable systems. In the 1990s a cable industry research consortium developed a standard for data communication over cable that has been embraced by most of the cable industry. This standardization permits many manufacturers to compete in the supply of compatible equipment and allows retailers to limit their inventories of cable modem equipment. It reduces or eliminates the risks that a consumer will be forced to invest in equipment with limited resale value or will be unable to use the modem after moving to a new city. The telephone industry does not have a similar uniform standard for DSL services.

Wireless and Satellite Services

U.S. households may also subscribe to a variety of wireless and satellite services. Unlike DSL and cable modems, most of these technologies will still take several years to develop fully and attract large numbers of subscribers.

Satellite systems are the most widely used of the high-speed wireless options today. Two services, DIRECWAY and StarBand, provide two-way satellite-based Internet access, using the two major direct broadcast satellite systems owned by Hughes and EchoStar. These systems provide service at data rates of about 400 to 500 kilobits per second and are principally attractive to households in rural areas that are not able to receive DSL or cable modem service.[7] The next generation of satellite systems may offer higher transmission rates and much more system capacity.

The Federal Communications Commission (FCC) has licensed two other radio services, multichannel multipoint distribution service (MMDS) and local multipoint distribution service (LMDS), which also can be used to provide wireless Internet access to fixed locations. MMDS was originally a wireless alternative to cable television. MMDS technology can serve customers within a range of about 15 miles of a base station and requires a line-of-sight path between the antenna at the residence and the antenna at the MMDS service provider's base station. Sprint and WorldCom have been the principal licensees of this spectrum, but both appear to have abandoned any attempt to use it to deliver residential broadband access.[8] LMDS operates at frequencies ten times higher than those used by MMDS, which means its transmissions are often impeded by rainfall.

Unlicensed radio bands provide yet another form of high-speed Internet access. A few hundred smaller Internet service providers (ISPs) have used this unlicensed spectrum to provide links to their customers. Typically, such systems transmit data at rates of about 1 megabit per second and can serve customers at ranges of up to 15 miles. Short-range communications using wireless local area network (LAN) technologies are much less likely to experience harmful interference than longer-range systems. This technology, popularly called WiFi, has spread very rapidly and has been deployed to provide broadband Internet access at airports, restaurants, and college campuses.[9] A variety of retail outlets, such as Starbucks, have deployed Wi-Fi in their facilities. Several airlines also provide wireless LAN connections in their airport lounges. These short-range services have not been introduced to homes or businesses yet, but they indicate the potential technological dynamism of broadband communications.[10]

In addition, the commercial wireless carriers, led by Verizon Wireless, are developing their own high-speed services. In 2003 Verizon Wireless launched a service that delivers 300–600 kilobits a second in Washington, D.C., and San Diego. In 2004 it announced that it would deploy this service in a large number of major metropolitan areas.[11]

The Diffusion of Broadband in the United States

The United States is in the fortunate position of having a highly developed cable television industry, a variety of (struggling) fixed wireless providers, and two large direct-to-home high-power satellite companies that can offer broadband Internet services to residences and small businesses in competition with its incumbent telephone companies. Indeed, because of the intense competition from the direct broadcast satellites launched by Hughes in 1994 and subsequently by EchoStar, cable television operators were forced to spend billions of dollars to upgrade their networks to deliver more programming options. In the process, these operators have added a two-way capability that allows most of them to offer cable modem service to their subscribers.

By the middle of 2004, the United States had 32.5 million "high-speed" lines in service, of which 30.1 million were provided to residences and small businesses (table 8-1).[12] Cable television systems accounted for 57 percent of these lines, and DSL continued to lag behind at 35 percent of the total. Cable modems have retained their lead despite an aggressive U.S. policy of promoting competition from new entrants offering DSL. Ironically, this policy is often blamed for the slow deployment of DSL because of the controversies it has spawned and its adverse investment incentives.[13] The United States also has a substantial number of fixed wireless and satellite operators offering broadband services. However, these fixed wireless and

Table 8-1. Growth of Broadband in the United States

Number of lines over 200 kilobits/second in one direction

Technology	*12/31/99*	*12/31/00*	*12/31/01*	*12/31/02*	*12/31/03*	*6/30/04*
ADSL	369,792	1,977,101	3,947,808	6,471,716	9,509,442	11,348,199
Other wireline	609,909	1,021,291	1,078,597	1,216,208	1,305,070	1,407,121
Coaxial cable	1,411,977	3,582,874	7,059,598	11,369,087	16,446,322	18,592,636
Fiber	312,204	376,203	494,199	548,471	602,197	} 1,060,502 (Fiber and Satellite or fixed wireless combined)
Satellite or fixed wireless	50,404	112,405	212,610	276,067	367,118	
Total	2,754,286	7,069,874	12,792,812	19,881,549	28,230,149	32,458,458

Source: FCC, *High-Speed Services for Internet Access: Status as of June 30, 2004* (December 2004).

satellite services accounted for less than 2 percent of all broadband lines as of the end of 2003.

As figure 8-3 shows, the United States was well ahead of the average European country in broadband penetration—broadband lines per 100 persons—at the end of 2003, but it lagged behind Korea, Canada, and several Benelux and Scandinavian countries.[14] The competitive U.S. DSL carriers (DLECs) account for less than 2 percent of all broadband lines, a share that is not expected to increase very much in the next five years (see figure 8-4). In Europe, the competitive DSL carriers' share, excluding resale, was 6 percent of all DSL lines at the end of 2003.[15] The only countries in which competitive DSL carriers capture a substantial share of broadband subscribers are Korea and Japan, but—as discussed in chapter 9—the regulatory regimes in these two countries differ substantially.

Because Korea, Canada, and a few other countries are now ahead of the United States in broadband penetration, there are repeated calls for public policy initiatives to accelerate the rollout of broadband by U.S. carriers.[16] However, rather than simply compare the current U.S. situation with that in other developed nations, it might be more useful to look at the diffusion rate for broadband alongside other revolutionary consumer technologies over the past fifty years. Figure 8-5 does this for television, cable television, the videocassette recorder (VCR), and the personal computer (PC). Note that broadband (shown by the heavy line) compares favorably with the VCR and the PC but lags behind television receivers. Thus the diffusion of broadband is not much different from the patterns of other notable technological breakthroughs. The case of cable television is particularly interesting because its growth was restrained by oppressive FCC regulation in the 1960s and 1970s. Whatever the deleterious impacts of regulation on broadband, they clearly have not been as bad as those imposed by the FCC on cable television.

Demand for Broadband Internet Connections

Hanging over the entire broadband debate is a concern that many households simply do not want the service at prices of $40 or more a month. Several years ago, Hal Varian and his colleagues at the University of California performed an experimental analysis of the demand for bandwidth in Internet connections.[17] Using members of the University of California community as subjects, they offered Internet connections at various speeds up to 128 kilobits per second at various prices. They found that

Figure 8-3. *Broadband Penetration, December 2003*

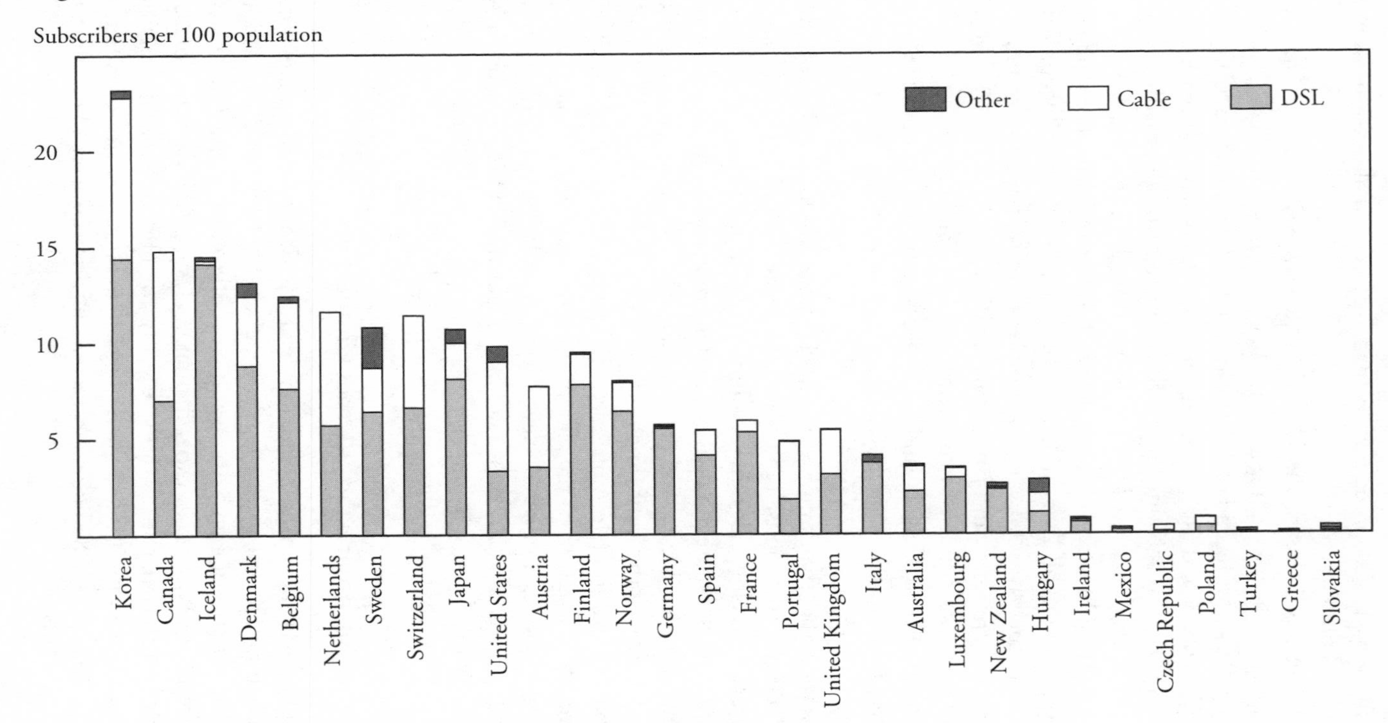

Source: OECD (www.oecd.org/document/31/0,2340,en_2649_34223_32248351_1_1_1_37409,00.html).

Figure 8-4. *U.S. Broadband: Forecast by Type of Carrier, 2002–08*

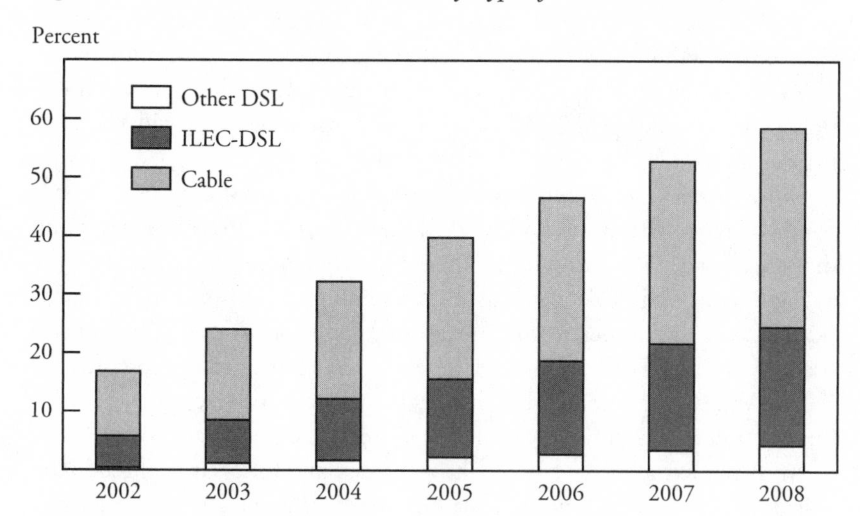

Source: Morgan Stanley (March 2004).

Figure 8-5. *Diffusion Rates for New Consumer Technologies*

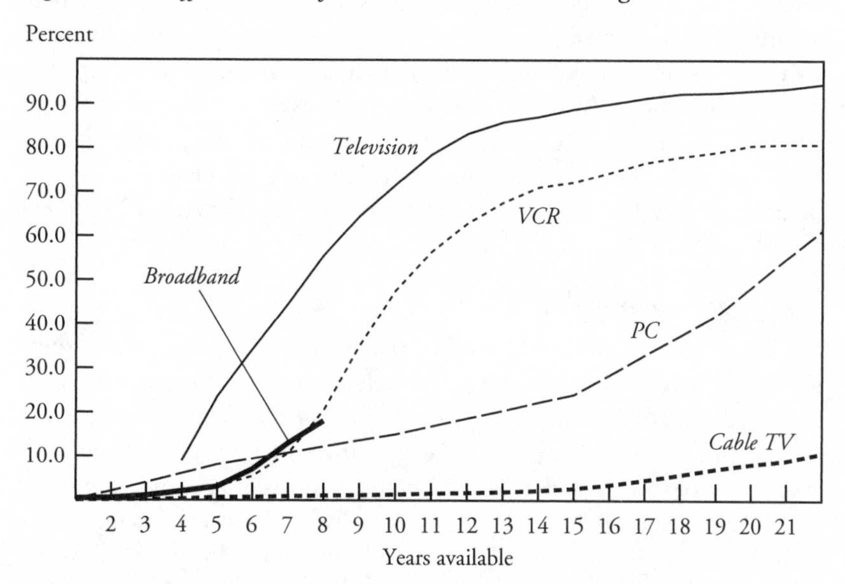

Source: Bruce Owen, "Broadband Mysteries," in *Broadband: Should We Regulate High-Speed Internet Access?* edited by Robert W. Crandall and James H. Alleman (Brookings, 2002), fig. 2.1, updated by the author.

few of their subjects were willing to pay more than $0.01 a minute, or less than $1 an hour, for the time saved in using the Internet. One reason for such a low willingness to pay is that the experiment was conducted in 1998–99, when most Internet users had little need for high-speed connections. This was before the MP3 craze and before widespread videostreaming developed.

More recently, in examining the nature of Internet use by U.S. broadband and narrowband subscribers, Paul Rappoport and his colleagues found that the average broadband user is on the Internet 50 percent more than the average narrowband user.[18] However, use among both groups is highly skewed, as is use of most communications media. Significantly, broadband users had more "visits" of Internet sites and were connected to those sites for shorter times—presumably because their connections were much faster—but the differences in the types of sites visited were not great.

At this juncture, no one knows how the demand for or use of broadband will unfold as new Internet applications are developed. At the same time, evidence from the United States suggests that the price elasticity of demand for broadband connections is fairly high. Rappoport and his colleagues found that the price elasticity of demand for DSL service is –1.46.[19] Using more recent data for households with access to both DSL and cable modem service in the United States, J. Gregory Sidak and I estimated that the price elasticities of demand for both cable modem service and DSL are equal to –1.2 and that the cross-price elasticities are positive.[20] These results are consistent with the observation that households still do not consider broadband a necessity. However, the limited evidence suggests that demand elasticities may be falling as U.S. broadband penetration approaches 25 percent of all households.[21]

Regulation of Broadband in the United States

Current discussions of U.S. public policy on broadband center on two issues: the need to regulate incumbent telephone companies that have alleged "bottleneck" monopolies and the effect of such regulation on these incumbents' willingness to undertake risky investments in broadband facilities. Much of this discussion focuses on the need to curb potential "monopoly power" even though the alleged monopolists—the local telephone companies—have only one-third of broadband subscribers. Indeed, there is still no evidence that any such power will develop given the competitive and technological struggle to deploy broadband services that now exists between cable companies, telephone carriers, and various wireless carriers.

This regulatory debate has raged for most of the nine years since the passage of the 1996 Telecommunications Act and is still not completely settled. As noted earlier, the incumbents' unbundling and line-sharing responsibilities under the act have been reduced as the result of the FCC's 2003 Triennial Review decision released in August 2003 (see chapter 2). Numerous other issues also remain before the FCC, and, as a result, the threat of new or expanded regulatory programs could reduce the incentive to invest in facilities and content at the dawn of the broadband revolution.[22]

Telephone Companies

For the most part, the U.S. Congress and the FCC have articulated a "hands-off-the-Internet" policy.[23] Broadband is an important exception to this policy. Incumbent telephone companies (ILECs) face a number of restrictions that have carried over from their legacy as "dominant" providers of ordinary telephone services.

Until recently, the FCC extended the unbundling requirements of the Telecommunications Act to elements required for broadband services. These obligations have applied not only to elements of the legacy network used to provide traditional voice services but also to parts of the network that are the result of new investments, which can be used to provide new services—even those that may not have been envisioned when the act was passed. Even though the FCC removed one of these responsibilities, "line sharing," in its August 2003 decision, the entire network unbundling regime remains a source of considerable uncertainty. This uncertainty is clearly an obstacle to the ILECs' decisions to upgrade their networks both to extend broadband services to more customers and to provide new services.[24]

The FCC and the states also require that the prices ILECs charge their end-user customers and ISPs be "just and reasonable," while other providers are free to set prices as market conditions permit or dictate.[25] Moreover, the ILECs must offer any services that they sell directly to end-users to their competitors at prescribed resale discounts. Given their role as regulated retailers and wholesalers of broadband services, the ILECs are also vulnerable to allegations of price squeezes or of unjust or unreasonable prices for the wholesale services that they offer to their competitors.[26]

Cable Companies

The 1996 act by and large deregulated cable television service prices and imposed no direct new requirements on cable providers that offer

Internet services, including cable modem services. As cable companies began to develop their cable modem broadband services, they designed these services to be funneled through their own proprietary ISPs, principally @Home and Roadrunner. As a result, independent ISPs began to petition municipal franchising authorities and regulators to require "open access" for all ISPs. However, there has been no regulatory attempt to require the cable companies to share their infrastructure with competitive broadband service providers.

After a series of court challenges, the issue of open access for cable systems was referred to the FCC because cable modem service is arguably a "telecommunications" service, not a cable television service. The FCC has yet to complete its rules on this question because its first ruling on the categorization of cable modem services as "information" services was overturned by the federal courts.[27] Therefore, cable modem service remains essentially unregulated, in contrast to the regulation still facing ILECs for their DSL services. This situation is generally referred to as "asymmetric regulation."

To date, only one of the rules applied to ILECs, the "open access" requirement that DSL services be available to multiple ISPs, has received any serious attention as being applicable to cable operators. Although this single obligation falls far short of the multiple obligations that ILECs have faced and that have surely affected their rollout plans, even the possibility of new obligations may be slowing cable modem deployment as well. As Thomas Hazlett has observed, cable operators have devoted only a small fraction (1/125) of their capacities to broadband access to the Internet despite the fact that the sharing of this limited capacity slows the cable modem service during hours of peak usage.[28]

"Layered Regulation," Vertical Integration, and Network Effects

In the United States, the debate over the regulation of broadband services has focused heavily on the need to provide new carriers with access to the incumbent telephone companies' infrastructure and the ability of Internet service providers to gain access to cable facilities. Little has been said about the network effects in the development of this new communications service.[29] The demand for high-speed connections depends on the availability of services (that is, "content") that require such high speeds. Until large numbers of households have broadband connections, however, the economic rewards for developing this content will be limited.

What is the solution to this "chicken-and-egg" problem? In earlier eras, such network externality problems were solved by vertical integration. The telephone service company (AT&T and its forebears) owned the telephone handset manufacturer. The motion picture companies owned theaters. Henry Ford integrated backward into parts development and manufacture and forward into vehicle retailing. Television networks initially produced their own programming. Cable television companies developed their own cable networks to fill a programming void as they expanded channel capacity.[30]

In the case of broadband, policymakers are focused on developing infrastructure competition without any concern for creating the complementary products required for this infrastructure to spread. Indeed, rather than encouraging the development of intellectual property through vertical integration, they often attempt to impede vertical integration through open-access requirements, separate subsidiary mandates, and tough merger standards. For instance, when AOL and Time Warner merged, the Federal Trade Commission required that the AOL-Time Warner cable systems provide access to ISP competitors of AOL.[31] Subsequently, the FCC imposed a number of conditions before approving the AOL-Time Warner merger:

> First, we find that the proposed merger would give AOL-Time Warner the ability and incentive to harm consumers in the residential high-speed Internet access services market by blocking unaffiliated ISPs' access to Time Warner cable facilities and by otherwise discriminating against unaffiliated ISPs in the rates, terms, and conditions of access. To remedy this harm, this *Order* conditions approval of the merger on certain conditions relating to AOL Time Warner's contracts and negotiations with unaffiliated ISPs. Second, we find that the merger would make it more likely that AOL Time Warner would be able to solidify its dominance in the high-speed access market by obtaining preferential carriage rights for AOL on the facilities of other cable operators. We particularly find that the merger would harm the public interest by allowing for *greater coordinated action* between AOL Time Warner and AT&T in the provision of residential high-speed Internet access services. To remedy these harms, we impose a condition forbidding the merged firm from entering into contracts with AT&T that would give AOL exclusive carriage or preferential terms, conditions and prices. Third, we find that the proposed merger would enable AOL Time Warner to dominate the next generation of

> advanced IM-based applications. To remedy this harm, we impose a condition requiring AOL Time Warner, before it may offer an advanced IM ("Instant-Messaging")-based application that includes streaming video, to provide interoperability between its NPD-based applications and those of other providers, or to show by clear and convincing evidence that circumstances have changed such that the public interest will no longer be served by an interoperability condition. (Emphasis added)

In short, both the Federal Trade Commission and the Federal Communications Commission perceived a potential threat from vertical integration in the AOL-Time Warner merger, and both imposed conditions designed to limit the potential harm from such integration. Unfortunately, these conditions also limited the benefits from AOL-Time Warner's investment in broadband content because AOL-Time Warner must provide other ISPs access to its cable modem customers and it must design certain content packages to be compatible with rivals' content. It is certainly possible that these conditions contributed to AOL-Time Warner's poor performance since the merger.[32]

Whether the restrictions on AOL-Time Warner and the regulatory requirements placed on the ISPs may eventually yield benefits in excess of their costs cannot be determined at this early stage of the development of broadband. However, the history of United States antitrust policy and regulation in matters involving new technologies and vertical integration does not provide grounds for optimism.[33]

This hostility to vertical integration in broadband is quite widespread, especially among lawyers and engineers who observed the birth of the Internet and its subsequent rapid growth. Many of these observers believe that the Internet was a success because of common standards and an "end-to-end" architecture that allows anyone to interconnect with essentially a static, "dumb" network using nonproprietary protocols. This reasoning derives from the observation that the Internet developed over the existing telecommunications network and that subscribers used existing voice-grade circuits to connect to it. Innovative applications, such as e-mail, could be developed to use the existing network and common, nonproprietary protocols, even by those who had no ownership interest in the network or contractual arrangements with its owners.

At the outset, however, subscribers generally accessed the Internet through simple dial-up connections over mainly local fixed-wire telephone systems under monopoly control. There were no cable modems, no 3-G wireless systems, no WiFi, and no fixed wireless systems available to would-be subscribers. Moreover, the fixed-wire Bell local telephone companies, over which subscribers connected to the Internet, were barred from communications between local access and transports areas (LATAs), making it impossible for them to offer Internet service. Because of these regulatory restrictions, they could not have redesigned the network and installed their own proprietary protocols to exploit opportunities for delivering their own applications or content over the Internet even if they had so desired. Until cable modem service emerged and the Bell companies were freed from the interLATA restrictions, no one could disrupt the end-to-end principle. By default, it was associated with the success of the Internet.

Surely it would have been better if the U.S. telecommunications network had been redesigned some time ago to provide higher speeds to users. Unfortunately, a widely diffused high-speed network did not exist in 1994 when the Internet began to grow rapidly, and there is no consensus even today on the most efficient design for such a broadband network or multiplicity of networks.[34]

Nevertheless, the end-to-end advocates continue to propose regulation of the network infrastructure, at least as far as preserving its common, open protocols and providing open access for all potential users of any network that develops. For example, in responding to the FCC's apparent intent to deregulate broadband, Mark Lemley and Lawrence Lessig state: "We do not yet know enough about the relationship between these architectural principles and the innovation of the Internet. But we should know enough to be skeptical of changes in its design. The strong presumption should be in favor of preserving architectural features that have produced this extraordinary innovation."[35]

Rather than deregulating broadband and allowing network owners to develop their own architecture, some would extend regulation on a "layered" basis, requiring the owners of the physical layer, that is, the network, to provide open access to all developers of applications and Internet service providers until there is demonstrated competition at the network level.[36] In order to provide such nondiscriminatory access, network owners would be required to continue to develop "dumb" networks that are accessible through standard, nonproprietary protocols. This would prevent network operators from developing their own applications and content and denying access to competitive applications and content. In an earlier era, this would have been similar to requiring Henry Ford, who had more than 60 percent of the U.S. automobile market in the early 1920s, from changing the design of his Ford platforms to adopt new technologies and deny his competitors the ability to install their components on his cars until GM and Chrysler became viable competitors.

The Effects of Regulation

Regulating a new service can generate large losses in economic welfare if such regulation increases the risk of investing in the facilities required to deliver the service. The consequent delay in the introduction of a new service or in the rate at which a new service is introduced denies consumers the opportunity to consume this service. In such cases, the economic costs to consumers can be quite high, particularly when the demand for these services is price inelastic. With inelastic demand, losses resulting from delay can be a large multiple of forgone revenues.[37] When the price elasticity of demand is high, consumers would be willing to pay little more than the current price for the service. As the demand for a service becomes less price elastic, however, the willingness to pay increases accordingly.

Charles Jackson and I have estimated that the benefits of universal broadband service in the United States could be as high as $300 billion a year to consumers and producers.[38] If broadband rollout is delayed by regulatory disincentives to invest, these gains—measured in terms of consumers' and producers' surplus—are likewise delayed. Even if the delay is just a few years, the present value of the losses to consumers and producers could be enormous, easily in the neighborhood of $500 billion.[39]

There is a growing perception that regulation has impeded the growth of the ILECs' DSL services. In 2001 one group of analysts downgraded its estimate of how fast ILECs would be able to roll out broadband services, explaining,

> While regulatory developments continue to favor cable [operators], the constraints on RBOCs [regional Bell operating companies] are increasing. Line sharing with other competitive local exchange carriers (CLECs) has been required for the Bells, resulting in increased competition for the RBOCs. Moreover, the establishment of separate subsidiaries for DSL operations has been required. As a result, some Bells are holding back on their aggressive rollouts, such as SBC in Illinois.[40]

There is little empirical evidence to support or disprove this hypothesis. Because so many other regulatory issues involving traditional voice services have been interwoven with broadband regulation, it is difficult to provide an empirical test of the effect of the regulatory environment on ILEC investment.[41]

In August 2003, when the FCC announced that it had decided to eliminate line sharing, it also voted to remove the unbundling obligation for investment in next-generation, fiber-optic networks.[42] It allowed incumbent LECs to extend new fiber to residences free of such regulation, while still requiring the ILECs to provide competitors with wholesale voice-grade services over such facilities. These decisions to end line sharing and reduce the threat of regulation for fiber to the home are surely a step in the right direction and may have had at least a temporary effect on ILEC investment (see figure 8-6).

In the case of broadband in the United States, the existence or threat of regulation has reduced the incentive to develop the necessary infrastructure to deliver broadband. The response of the Bell companies has been to slash capital spending, which had been rising steadily in 1998–2000 (see figure 8-6). Capital spending per line in 2002 and the next few years had been projected to be far below even its pre-1996 level despite the considerable capital requirements to complete the rollout of DSL, which had been stalled at about two-thirds of U.S. residential and small business locations (figure 8-6).[43] However, estimates published after the FCC's February 2003 decision to end line-sharing regulation suggest a rebound in the incumbents' capital spending plans. The downturn in incumbent capital outlays may have been arrested by the FCC's apparent move toward modest deregulation. Similarly, forecasts of ILEC rollout of DSL, measured in terms of "addressable locations," also increased after the FCC decision (figure 8-7).

Using Regulation to Promote Broadband Competition

The new local entrants that specialize in offering DSL services over incumbent loops have fared very badly. Only one of the three largest of these new carriers—Covad—remains in operation after its reorganization in bankruptcy. A few others, such as 360Networks and DSL Net, also survive, but their prospects seem bleak. Covad reported that it had 517,000 subscribers at the end of 2003, but its future growth is now in doubt because it relied heavily on those entrants (AT&T and MCI) that provided local service through the UNE-P. Given the court decision in 2004 that reversed the FCC's attempt to keep UNE-P alive, Covad's prospects do not look good.[44]

Figure 8-6. *ILEC and Cable TV Companies' Capital Expenditures, 1996–2006*

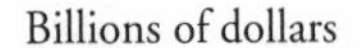

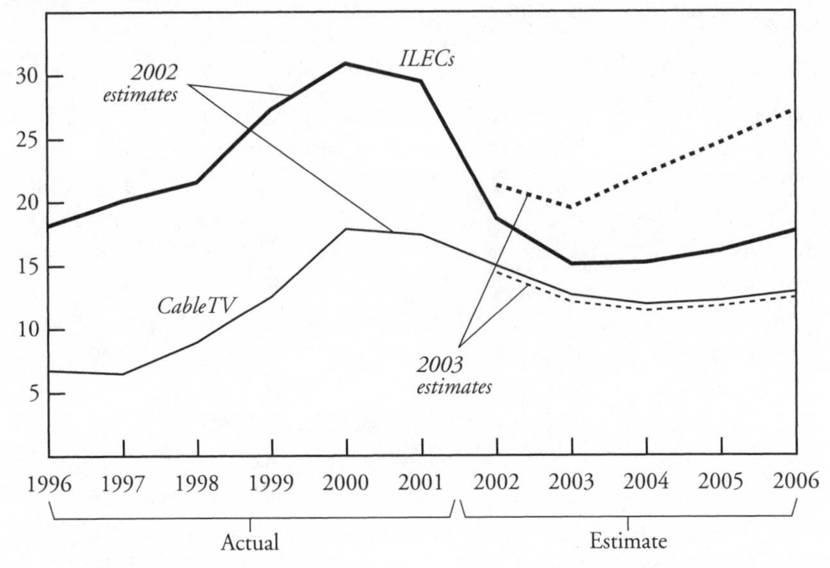

Source: CreditSuisse-First Boston (2002, 2003).

There is no evidence that network sharing has increased competition in U.S. broadband markets. At the end of 2003, the FCC reported that only 1.7 percent of all broadband lines were DSL lines offered by nonincumbent telephone companies. Another 1.5 percent reflected nonincumbent telephone companies offering a wireline service other than ADSL or cable modem service.[45] Thus the share of broadband lines accounted for by entrants, in toto, was only 3 percent, and the share using the incumbent telephone companies' facilities remains mired in the range of 1 to 2 percent.

In a recent analysis by Debra Aron and David Burnstein, facilities-based competition between cable modems and incumbent telephone companies' DSL services has a statistically significant positive effect on overall broadband penetration in the United States, but the price of UNE loops has no statistically significant effect.[46] This is further evidence of the futility of a regulatory policy focused on making incumbent telephone companies' facilities available to competitors rather than opening more lines to potential final consumers.

Figure 8-7. *Forecast of Addressable Broadband Locations, 2002 versus 2003*

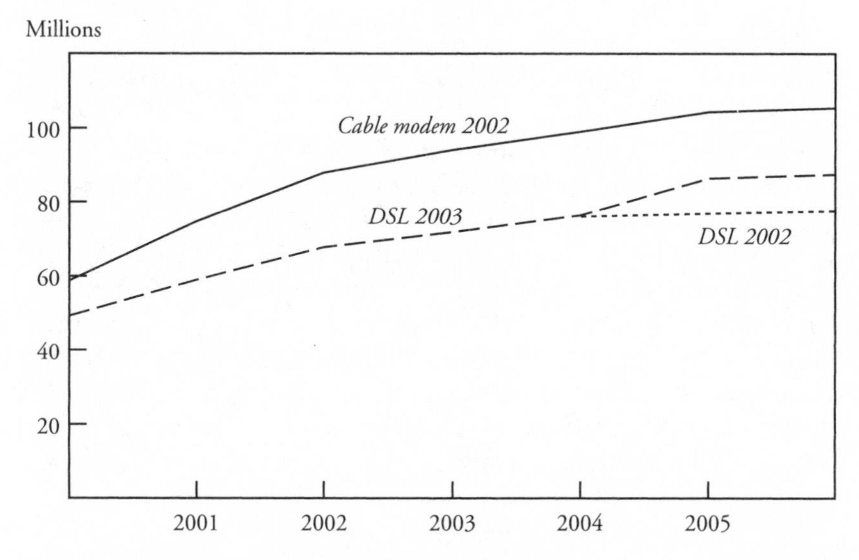

Source: Morgan Stanley (October 25, 2002, and April 6, 2003).

As Michelle Kosmidis reports, unbundling has not worked any better in Europe.[47] Although the European Union has avoided the U.S. mistake of using unbundling to promote local entry into voice services, its more limited local-loop unbundling policy to promote broadband appears to be failing. At the end of 2003, only 6 percent of all DSL lines in Europe were delivered by CLECs over unbundled or shared ILEC loops.[48]

Toward Truly High-Speed Connections

The regulatory controversy over wholesale unbundling and line sharing in the United States has obscured an important fact: U.S. "high-speed" connections are not very fast. A speed of 600 kilobits a second or even 1 megabit a second may seem a great deal faster than the speed of a subscriber's old dial-up modem, but the technology is much further along than this. Broadband service in Korea and Japan routinely performs at 8 megabits a second, or even higher. With a modern infrastructure and densely packed residential areas, these two countries are now sprinting ahead of the United States both in deployment and in quality of service. By the time the United States succeeds in diffusing broadband to 30 or 40 percent of the population, these subscribers will probably be just as dissatisfied with their "high-speed" service as an earlier generation was with the dial-up worldwide-wait. What is the solution to this problem? At present, all signs point to direct fiber connections, or "fiber to the home" (FTTH), or at least fiber very close to the home.

In Japan and Korea, fiber-optic capacity has been extended very close to the final subscriber. New apartment buildings in Korea are being wired with fiber optics that can deliver speeds of up to 100 megabits a second, and in Japan, the incumbent telephone company (NTT) is extending fiber to residences in densely populated areas (see the discussion of Japan and Korea in chapter 9). In the United States, such an investment program is now beginning, perhaps in response to the FCC's 2003 decision to end line sharing and reduce the wholesale regulation of fiber to the home.

Why would households be interested in such high speeds if only 15–20 percent of them now subscribe to broadband? The answer lies in the applications that high speeds can handle (see table 8-2). Needless to say, it will be some time before most Americans can download video in real time from their current broadband connections, but many Koreans already can.

The cost of deploying fiber to the home is declining rapidly. The most likely approach is to extend fibers out in the network to a point where they are "split" into eight, thirty-two, or more distribution fibers to the subscriber. This approach economizes on the use of fiber in the network and reduces connection costs at the central office or headend of the provider. The most expensive part of a direct fiber-to-the-home architecture has been the installation of the equipment required at the customer's premises. An optical network unit is needed to convert the light wave into the electrical pulses that household telephones and television sets now use. The customer may also have to install a source of standby power since the fiber connection cannot conduct electricity. Like most electronic devices, such equipment is going down in price, which may now be as low as $100 to $200 per subscriber, but installation costs are another story: recent estimates put the full cost of installing FTTH in urban areas at as much as $1,100–$1,400 per subscriber.[49]

More and more large, incumbent U.S. telephone companies are beginning to explore the possibility of extending fiber to the home. SBC has tested its FTTH system in Mission Bay, a district of San Francisco.[50] BellSouth began testing its FTTH system in Atlanta and Fort Lauderdale in 2001. At present, Verizon has the most ambitious plans to roll out FTTH

Table 8-2. *Speeds Required for Various Activities*
Megabits/second

Activity	*Speed*
Interactive gaming	1.0
Video conferencing	1.0
Web surfing	1.5
Telecommuting	2.0
Videostreaming	15.0
High-definition television	19.0

Source: Dan Hickey, "Jump into the PON," *Communications and Engineering Design* (November 2002).

by spending $2.5 billion to pass 3 million homes by the end of 2005.[51] BellSouth, Qwest, and SBC are pursuing a less aggressive strategy of extending fiber to neighborhood nodes. Despite this growing interest, FTTH technology had spread to fewer than 20,000 homes and small businesses by the end of 2003.[52]

The competitive carriers now co-located in ILEC switching centers have designed their equipment to use the copper loops of the incumbents. If fiber optics replaces the copper loops, will the incumbents eventually have to provide the entire fiber path, including central-office (ATM) switching facilities to entrants at forward-looking costs? If so, the benefits of risking $1,100 to $1,400 per home are surely attenuated.

Conclusion

With its extensive cable television penetration and household Internet use, the United States is poised to enjoy substantial benefits from aggressive platform competition between cable operators and telephone companies for broadband customers. This competition is augmented by smaller niche players offering broadband services via satellite and fixed wireless platforms. There are now about 30 million residential broadband subscribers, and the largely unregulated cable modem services have nearly 60 percent of them. Given the substantial price elasticity of demand for such services, technical progress and the resulting decline in the price of broadband services will likely provide further impetus for household subscriptions.

Unfortunately, most of the past nine years have been spent debating how to "open up" the telephone companies' networks to competitors incapable of building or unwilling to build their own facilities and how to provide ISPs with access to cable platforms. This exercise has been costly and largely unsuccessful. Most of the new local entrants are either mired in bankruptcy or nearly so, and the smaller ISPs are disappearing anyway. Equally important, the incumbent Bell companies slashed capital expenditures between 2000 and 2002, despite the fact that about 30 percent of households and small businesses were still unable to receive DSL service.

In 2003 the FCC decided to sharply reduce the degree of line sharing and unbundling required of incumbent telephone companies. This decision appears to have reversed the decline in Bell-company planned capital expenditures, at least temporarily, and has led the Bell companies to reduce DSL prices and roll out service to heretofore unserved areas. In turn, cable television systems have responded by increasing the speed of their cable modem services. Broadband subscriptions have been increasing rapidly and may now accelerate in response to the lower prices. Verizon is even beginning to deploy fiber to the home, but uncertainty over future FCC rules and declining revenues may still impede such investments. At this early stage of development, it is difficult to see precisely how the speeds available from direct fiber connections will be used by households. Nevertheless, the continuing diffusion of broadband at greater speeds throughout the United States and elsewhere will undoubtedly spark further innovation in broadband content.

9 Telecom Reform in Other Countries

Telecom liberalization has now spread to virtually every developed country in the world and even many developing countries. Although the United States has had perhaps the most aggressive policy of encouraging local entry and has had long-distance competition for several decades, other countries have also opened their telecom markets to competition in recent years. The United Kingdom began by privatizing British Telecom in 1984 and opening entry into telecom services in 1985.[1] Japan opened its national and international calling markets to competition in 1985, and Canada liberalized its long-distance market in 1992 and its local telecom markets in 1997.[2] The European Union's Telecommunications Directive opened all EU telecommunications markets to competition in January 1998.[3]

Canada

Canada was much slower than the United States in opening its telecom markets to competition. Several telephone companies were owned by provincial governments, and the federal government did not have regulatory authority over the entire country until 1992.

Regulatory Policy

Canada admitted facilities-based entrants into long distance at the end of 1992 and required the incumbent carriers to provide equal access (that is, equal quality connections) to all long-distance carriers. Incumbent telephone companies retained their integrated operations and were allowed to respond to entry by lowering their retail rates.

In 1997 the Canadian Radio-Television and Telecommunications Commission (CRTC) opened local telecommunications markets to competition by requiring incumbents to interconnect with new entrants, provide "essential services" for resale, and lease unbundled network elements.[4] The CRTC did not require a mandatory discount for the offering of services at wholesale prices, and unbundled elements were to be leased at estimated cost plus a 25 percent markup. Unlike the U.S. Federal Communications Commission (FCC), the CRTC extended unbundling only to "essential network facilities," such as local loops in suburban and rural areas, central office codes, and subscriber listings. Loops in urban areas, interoffice transit (transmission), and electronic signaling were not deemed essential facilities and were therefore to be unbundled for only an "interim" period of five years. Local switching was not unbundled at all, even over the five-year transition period.

Because local competition has been slow to develop in Canada, the CRTC was subsequently persuaded to extend the deadline for the "interim" requirement of local loop unbundling in urban areas. In addition, it recently reduced the markup on unbundled element costs from 25 percent to 15 percent, but it rejected a proposal by AT&T Canada to emulate the U.S. UNE-P requirement by forcing the incumbents to offer their entire service platform at a 70 percent discount from retail.[5] Although the CRTC has not mandated line sharing, the incumbents, led by Bell Canada, now offer line sharing to digital subscriber line (DSL) entrants at rates of $5 a month per line and even less.

By allowing integrated incumbent carriers to compete aggressively with the new entrants and by limiting the entrants' access to incumbent facilities at low, wholesale rates, the CRTC has been much less aggressive in promoting or even "subsidizing" entry than has its U.S. counterpart. This more conservative approach has allowed competition to develop at a more modest rate without creating the stock market frenzy that gripped the United States.

The Results of Liberalization: Narrowband

Competition in long-distance services increased rapidly after 1992. By 2002 the new entrants, principally Call-Net (Sprint) and Allstream (AT&T Canada), accounted for 30 percent of long-distance revenues (see

Table 9-1. *Telecom Competition in Canada, Entrants' Share, 2003*
Percent

Category	*Business*	*Residential*	*Total*
Long-distance minutes	45.9	27.1	36.5
Long-distance revenues	41.5	23.6	30.3
Local lines	11.7	2.0	5.4
Local revenues	7.9	1.9	4.3
Internet access service revenues	43.5	14.8	22.1

Source: Canadian Radio-Television and Telecommunications Commission, *The Status of Competition in Canadian Telecommunications Markets,* Report to the Governor in Council (November 2004).

table 9-1).[6] The result was a rapid decline in the average price of long-distance services to levels that are very close to U.S. rates or perhaps slightly lower, despite the fact that competition in U.S. long-distance services began almost twenty years before Canada liberalized its market.[7] By contrast, local competition in Canada has grown very slowly. In 2003 the new competitors accounted for 5 percent of access lines and only 4 percent of revenues, most of it coming from business service.

Industry Canada listed 117 "alternate providers of long distance," 497 resellers, and just 20 competitive local exchange carriers (CLECs) in 2002, but only three major national entrants—Call-Net, Allstream (formerly AT&T Canada), and GT Group—have been able to capture much market share.[8] Call-Net and Allstream offer long-distance and local voice services as well as data services. GT is primarily an Internet/data services provider. Each of these companies has encountered severe financial problems.

Call-Net had a market capitalization of less than US$140 million at the end of 2003 and outstanding long-term debt of US$327 million. Its stock fell by more than 90 percent after the introduction of local competition in 1997, and it has recently gone through a major financial reorganization.[9] AT&T Canada had been a subsidiary of AT&T until recently, but it too was reorganized financially because of continuing losses. In October 2002, AT&T negotiated with the holders of Can$4.5 billion of AT&T Canada's debt to accept a 69 percent interest in exchange for the debt. According to AT&T's quarterly report to the Securities and Exchange Commission in November 2002, this transaction "cost" AT&T upward of $2.5 billion.[10] AT&T Canada was reorganized as Allstream in June 2003 and was acquired by Manitoba Telecom Services, the Manitoba incumbent, in 2004 for

Can$1.2 billion. GT also encountered severe economic problems. After several years of losses, it was sold to 360Networks in 2002 for US$168 million.[11] In 2004 Bell Canada—the largest incumbent telephone company in Canada—acquired it from 360Networks.[12] Thus the market value of the three major competitive Canadian carriers, before MTS's acquisition of Allstream, had declined to less than US$1.5 billion. This compares with approximately $44 billion in market value for Canada's incumbent carriers.[13]

Competition in Broadband Services

As just mentioned, Canada's regulators have historically been much less open to unlimited unbundling than their counterparts in the United States. Unbundling was not extended to broadband providers through line sharing until October 2000, and there has been little apparent interest in it on the part of the entrants who use the incumbents' loops. However, as in the United States, significant broadband competition has come from cable. Canada has a well-developed cable television industry, with approximately 68 percent of households subscribing to cable television service, among the highest cable subscriber penetration in the world.

Overall, Canadian broadband penetration is about 40 percent higher than in the United States (see figure 8-3), and approximately 57 percent of Canada's broadband subscribers use cable modem service. The incumbent telephone companies account for virtually all of the DSL service, and they serve a substantially larger share of the population with DSL than the incumbent local exchange carriers (ILECs) do in the United States. TELUS and Bell Canada charge only US$27 for a DSL service of 1.0 to 1.5 megabits per second, far less than the $35 to $40 generally found in the United States.[14] Cable modem service is available from the largest Canadian cable company for US$34 a month.[15] The much less intrusive Canadian regulation appears to be working.

Thus more than ten years after the CRTC began to admit facilities-based entrants into Canadian telecommunications, the new local entrants constitute only a small share of Canada's narrowband market. Although they have not been able to develop profitable business plans, they have succeeded in driving down Canada's long-distance rates. Furthermore, competition in broadband is at least as robust as in the United States, and it has developed without the extensive unbundling regime that U.S. regulators erected and subsequently dismantled. On the whole, the CRTC's conservative regulatory policy has been remarkably successful, and it has allowed Canada to avoid the excesses of an investment boom followed by a collapse in stock market values—that is, it did not experience a "bubble" of the magnitude of the one that engulfed telecommunications in the United States.

The European Union

The European Union began liberalizing its telecommunications market on January 1, 1998. Europe had much more work to do than its counterparts in North America because most European telecom sectors had been dominated by a single, government-owned national operator until the late 1990s.

The Regulatory Choices

Under the EU liberalization, all countries were required to privatize their incumbent telephone companies, open these companies' networks to competition, and establish an "independent" regulatory authority. Although the broad dimensions of EU regulation are set by the European Commission in Brussels, the details of formulating regulations and enforcing them are left to the national regulatory authorities. In 1997–98 the European Union released a series of directives that would guide the liberalization of telecommunications in member countries. In accordance with these directives, each country's independent regulatory authority must admit new entrants, provide carrier pre-selection and number portability, set prices for interconnection between incumbents and entrants, limit "universal service" subsidies, and initiate a variety of other regulations governing incumbents and entrants.[16] Although most countries were to begin opening their sectors on January 1, 1998, there was no express policy for unbundling network facilities until December 2000.[17] In the interim, entrants could build their own facilities or obtain access to customers by using the incumbents' services.

The Results: Narrowband

Competition in the first three and one-half years focused almost exclusively on *calling*, not access. To obtain access to customers, entrants could use the incumbents' networks to originate and terminate their calls, install some of their own switches and transport facilities, or build their own facilities. As in the United States and Canada, facilities-based entry has been slow. From the share of subscribers served "directly" (that is, over the entrants' own lines) and the share of revenues obtained by entrants from direct and "indirect" subscribers (those served over incumbents' lines), it is

Table 9-2. *Entrants' Share of Fixed-Telephony Markets in EU Countries, 2001–02*
Percent

Country	*Share of subscribers served "directly"*	*Share of total fixed-telephony revenues*
Austria	10	11
Belgium	9	23
Denmark	. . .	8
Finland	0	0
France	7	20
Germany	1	20
Ireland	15	22
Italy	0	24
Netherlands	11	24
Norway	12	17
Portugal	0	5
Spain	4	14
Sweden	13	45
United Kingdom	2	26

Source: IDC, *Monitoring European Telecoms Operators: Final Report* (January 2002). Report filed with the European Commission (http://europa.eu.int/information_society/topics/telecoms/implementation/index_en.htm).

evident that competition varies widely across countries (see table 9-2). Ireland, the Netherlands, Norway, and Sweden have the largest estimated share of lines served by entrants through "direct" competition.

Because telephone companies in all EU countries charge customers for local calls, competition has focused largely on offering lower prices for both local and long-distance (national and international) calls. Long-distance competition has been much more intense than local competition because the incumbents' margins on national and international calls have traditionally been used to subsidize local connections. According to somewhat different data published by the European Commission, the entrants' share of national and international calling revenues was more than 30 percent in more than half of the EU countries by the end of 2002 (see table 9-3). By contrast, competitors' shares of local calling revenues exceeded 30 percent in only one country, Austria.

Table 9-3. Entrants' Share of EU Telecommunications, December 2002
Percent

Country	*Long-distance minutes*	*Long-distance retail revenues*	*International minutes revenues*	*International retail minutes*	*Local revenues*	*Local calling*
Austria	55.0	n.a.	59.0	n.a.	34.0	n.a.
Belgium	18.7[a]	17.1[a]	44.3	38.4	18.7[a]	17.1[a]
Denmark	32.6[a]	n.a.	45.9	n.a.	32.6[a]	n.a.
Finland	55.4	38.4	45.1	32.0	n.a.	17.0
France	40.4	30.1	36.6	31.9	24.4	20.8
Germany	40.0	35.0	58.0	41.0	8.0	4.6
Greece	5.0	2.6	4.9	3.9	0.6	0.7
Ireland	40.4	31.0	36.6	26.0	24.4	5.0
Italy	n.a.	30.8	n.a.	35.5	n.a.	23.3
Luxembourg	20.0[a]	n.a.	22.0	n.a.	20.0[a]	n.a.
Netherlands	n.a.	25.0	n.a.	35.0	n.a.	10.0
Portugal	9.1	n.a.	23.0	n.a.	n.a.	n.a.
Spain	19.8	23.4	38.7	35.1	24.3	18.3
Sweden	n.a.	42.0[a]	n.a.	57.0	n.a.	42.0[a]
United Kingdom	48.1	37.3	68.0	48.9	27.1	38.0

Source: European Commission, *Ninth Report on the Implementation of the Telecommunications Regulatory Package* (2003), annex 1, figs. 7 and 8.

n.a. Not available.

a. Local and long-distance calls.

The European Commission data do not include market shares for local line rentals, that is, local connections. According to the commission, very little of this local competition in the European Union occurred over the entrants' own facilities.[18] Of the twelve countries for which data are available, only six indicate that at least 5 percent of subscribers used an alternative facilities-based operator for local access in 2003. Thus local entry in most of the European Union has been based on resale of the incumbents' local services to provide lower-cost local calls.

Unbundling was not mandated in Europe until December 2000 and was designed to promote competition in broadband Internet services (DSL), not voice services. Of 193 million incumbent lines in EU member countries, 2.9 million had been leased as unbundled loops or as shared lines through December 2003.[19] Of these unbundled loops, 1.4 million are used for narrowband services, and 950,000 of these are in Germany.[20]

The Results: Broadband

As in the United States and Canada, broadband competition has developed across platforms—between cable systems and incumbent telephone companies. Broadband penetration is generally greater in countries with a well-developed, independent cable television sector. Those with the greatest broadband penetration, such as Belgium, Denmark, Austria, and the Netherlands, have a substantial number of cable modem connections (see figure 9-1). For this reason, the European Commission's policy has concentrated on encouraging the separation of cable television ownership from the ownership of incumbent telephone companies and forcing local loop unbundling. Neither has been very successful. Germany's attempt to force Deutsche Telekom to divest itself of its extensive cable assets was slowed for a variety of reasons. And unbundling simply has not led to a vibrant, independent DSL sector, as figure 9-1 shows. Only Denmark, Finland, and the Netherlands have any measurable unbundling or line sharing for independent DSL operators. For the entire European Union, these companies accounted for only 1.4 million lines out of a total of 22.5 million broadband connections in December 2003.[21] Europe also allows entrants to resell the incumbents' DSL facilities through "bitstream access" to the incumbents' DSL services. This bitstream access and simple resale accounted for another 3.2 million lines.[22] Thus, even by the most expansive definition of competition from nonincumbent providers of DSL, these new entrants accounted for just 4.6 million lines in December 2003, or slightly more than 20 percent of the total broadband market, with 70 percent of it through resale.

The New Entrants

As in the United States, the new entrants into the telecommunications sector in Europe are struggling. By 2001 entrants had obtained somewhat less than 20 percent of telecom revenues (table 9-2), but these entrants reached only a small share of their subscribers directly over their own facilities. Most of the entrants therefore offer calling services over the incumbents' lines, exploiting arbitrage opportunities between regulated wholesale rates and retail rates.

The EU entrants that have built their own networks have fared badly, as can be seen from the decline in stock market capitalization of the major public entrants since January 2000 (table 9-4). Of the major entrants in Europe, only Tele2 was able to retain at least half of its January 1, 2000,

Figure 9-1. *Sources of Broadband in the European Union, December 2003*

Broadband lines per 100 telephone lines

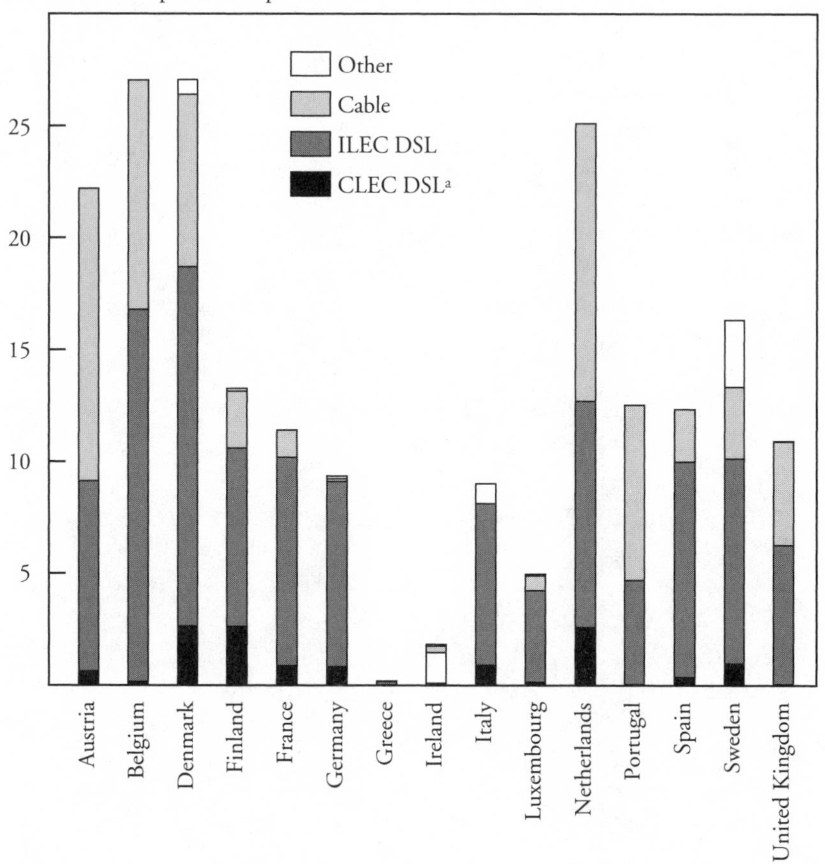

Source: ECTA, *Scorecard* (December 2003).

a. Reflects lines served through unbundled loops or line sharing with the incumbents. The figure does not include the resale of the incumbents' broadband services.

market cap into 2003. Tele2 offers traditional voice telephony services, wireless (cellular) service, cable television, and Internet services over its own networks in twenty-two European countries. In the first quarter of 2003, its total customer base increased to 17.7 million subscribers. It reported increasing revenues and was profitable in 2002–03.[23] Since then, however, even Tele2 has declined severely.

Table 9-4. *Market Capitalization of Alternative Facilities-Based Network Providers in Europe, 2000–04*

Millions of U.S. dollars

Date	*Colt*	*Energis*	*Fibernet*	*Thus*	*Completel*	*Equant*	*Jazztel*	*Song Networks*	*Tele2*	*Utfors*	*Versatel*
2000											
January 1	34,320	8,401	1306	4,444		23,217	3,890		7,038		2,686
June 30	23,263	6,404	1285	2,958	1,962	8,220	1,474	1,907	7,655	761	3,450
2001											
January 1	15,048	219	696	603	542	5,268	706	761	5,888	351	730
June 29	4,867	4,959	355	397	400	5,074	355	330	4,620	150	250
2002											
January	2,500	1,425	421	534	166	3,381	302	142	5,215	99	85
June 28	988	45	45	247	32	1,691	106	31	2,692	22	25
2003											
January 1	1,104	7	20	195	10	1,146	6	18	3,887	9	28
2004											
June 8	2,371	n.a	132	668	215	1,955	107	327	936	n.a.	876

Source: Datastream, as reported in Robert W. Crandall and Leonard Waverman, "Entry Strategies in Landline Markets," paper prepared for France Telecom (2003).

n.a. Not available.

As table 9-4 shows, there have been few apparent successes among European entrants. Cable & Wireless was once British Telecom's principal competitor in traditional mass-market voice telephony in the United Kingdom, but it abandoned this business to focus on developing a global high-speed network offering advanced services for business customers, a venture in which it has performed very badly. Equant is a relatively new carrier offering traditional and advanced telecom services to business customers, but it is controlled by France Telecom, which owns 54 percent of its stock. It too has lost about 95 percent of its market capitalization in the past three years.

Japan

Like most other developed countries, Japan has begun to privatize its national telecommunications carrier, Nippon Telephone and Telegraph Corporation (NTT), and has opened its telecommunications markets to competition. The process started in 1985, and then in 1996 NTT, the incumbent carrier, was asked to restructure itself into a holding company with a separate long-distance division and two local operating companies, NTT East and NTT West. By March 2002, the new entrants, or new common carriers, had obtained 25 to 28 percent of the long-distance market but only 20 percent of the local market.[24]

In December 2000, the Ministry of Public Management, Home Affairs, Posts and Telecommunications introduced a network unbundling requirement in Japan, allowing new entrants to offer DSL services over shared NTT lines. At that time, Japan had fewer than 1 million broadband subscribers, most of them using cable modem service. Since that time, DSL has grown rapidly, attracting 11.5 million subscribers by March 2004, while cable modems had grown to just 2.6 million lines and fiber to the home had increased to 1.2 million lines.[25]

By March 2004, the new entrants had accounted for nearly 62.5 percent of Japanese DSL subscriber lines, thanks largely to aggressive price competition from Yahoo-Broadband (Yahoo BB), which is offering DSL for as little as ¥2,280 (about $20) a month, not including the cost of the modem.[26] In its first year, Yahoo BB had more than 1 million subscriber lines, and by March 2004, it had increased its subscribers to 4.28 million using lines shared with NTT.[27]

Yahoo BB is a subsidiary of Softbank, which provides the financing and infrastructure for Yahoo BB's operations. Although Yahoo BB has reported substantial profits from its operations, Softbank has continued to report very large losses in its Broadband Infrastructure Division.[28] In the most recent fiscal year, ending March 31, 2004, it reported an operating loss of ¥87.2 billion on sales of ¥128.9 billion. Given an average of 3.3 million subscribers in the fiscal year and exchange rate of ¥110 to the U.S. dollar, Softbank realized $355 per year in revenues per subscriber, but it lost $240 per subscriber before interest and taxes.[29] These losses follow an even larger loss of $96 billion in fiscal year 2003.[30]

Softbank's objective is to build a very large customer base to which it can sell a variety of entertainment and information services, as well as voice over Internet protocol (VoIP). It has been extremely successful in selling VoIP to its subscribers, with 4.04 million of its 4.28 million DSL subscribers accepting the service.[31] Whether its strategy of building market share and eventually selling enough content to offset its huge start-up losses can succeed, no one can know at this time, but there is surely reason to fear that it is following even more aggressively in the steps of the U.S. trio that went bankrupt in 2000–01: Northpoint, Rhythms, and Covad.

The only countries with a substantial number of DSL lines supplied by new entrants are Korea and Japan. Since Korea did not have a mandatory unbundling policy until recently, only Japan offers evidence that new entrants using unbundled loops are contributing significantly to the diffusion of broadband Internet access services. Whether this "success" can be sustained in light of Softbank's enormous losses remains to be seen.

The network unbundling policy has not been used extensively for new entry into narrowband voice services in Japan until recently. In 2004, however, Softbank purchased Japan Telecom, a carrier that provides telecom services principally to larger corporate subscribers. Subsequently, Softbank announced that it would expand Japan Telecom's operations into the residential market by leasing lines from NTT and offering service at a very small markup over the wholesale price of the NTT lines. Given the results in the United States for entrants who adopted such a strategy and Softbank's demonstrated success in attacking the residential voice market through VoIP delivered over Yahoo-BB's DSL lines, Softbank's announced strategy is puzzling at best.[32]

Korea

Although Korea's telecommunication sector has been opened to competition, the government has been more concerned with developing a high-speed infrastructure than with establishing competition in traditional narrowband services. In 1993 it introduced a plan for the national "information infrastructure." Two years later, it published a three-stage plan to develop a high-speed infrastructure with government funding, and by 2000 it had completed a nationwide backbone of optical-fiber and ATM switches.[33] Much of this backbone was built by the country's electric utility company, but it is available to all telecom carriers who wish to offer high-speed services. In addition, all new apartment buildings have built advanced high-speed distribution capabilities into their buildings, including fiber optics and even DSLAMs.[34]

The incumbent telephone company in Korea, Korea Telecom, continues to dominate the traditional telephony services. As of February 2003, Korea Telecom had 96 percent of local telephone lines, including ISDN lines.[35]

The principal competition in the Korean telecom sector is found in the broadband services market. The three major competitors are Korea Telecom, Hanaro, and Thrunet, which account for 86 percent of broadband subscribers, but there are several other competitors (see table 9-5). Each major competitor has built a substantial infrastructure, and each has access to the electric utility's infrastructure.[36] Indeed, between 2000 and 2002, Hanaro, Thrunet, and Dacom invested a combined $5.7 billion in network assets, or almost as much as Korea Telecom.[37] The result has been aggressive facilities-based competition among these companies.

This competition has developed without government regulators requiring the incumbent to unbundle its facilities. The Korean government waited until December 2001 to announce a network unbundling policy because it wanted to encourage investment. The Ministry of Information and Communications acknowledged this strategy in its 2002 information technology report: "It is noteworthy that Korea enacted its statutory and regulatory requirements for incumbent facilities-based suppliers (KT, Hanaro, and Thrunet) to provide unbundled network elements to their competitors when the country was well on its way to leading the world in broadband access penetration rates."[38]

Korea had twenty-five broadband lines per 100 persons in December 2003.[39] By comparison, the United States had just under ten broadband lines per 100 persons, and the European Union had less than six. Clearly, Korea's policy of encouraging broadband facilities deployment, in part through government subsidies, has worked, but it is not clear how much of the difference between Korea and the rest of the world is due to these policies and how much should be attributed to differences in consumer tastes.[40]

Table 9-5. *Broadband Competition in Korea, December 2003*
Number of subscribers

Carrier	*DSL*	*Cable modem*	*LAN in apartment building*	*Satellite*	*Total*
Korea Telecom	5,230,342		353,580	4,836	5,589,058
Hanaro	1,093,261	1,290,190	342,152	. . .	2,725,563
Thrunet		1,287,502	5,862	. . .	1,293,364
Onse		419,293	3,769	. . .	423,082
Dreamline	56,178	89,546	3,874	. . .	149,598
Dacom		135,884	65,820	. . .	201,704
Value-added carriers	3,382	605,791	9,950	. . .	619,103
Resellers	52,812		124,235	. . .	177,047
Total	6,435,955	3,828,166	909,542	4,836	11,178,499

Source: Korea Ministry of Information and Communication, *White Paper Internet Korea 2004* (www.mic.go.kr/index.jsp [August 19, 2004]).

Comparing the United States and Other OECD Countries

Some evidence on the comparative effects of competition may be found in telephone prices and investment in communications equipment in the United States and other countries of the Organization for Economic Cooperation and Development (OECD). It is important to stress, however, that price comparisons may be misleading because of continuing regulation of rates and service offerings in most countries.

Liberalization and Telephone Prices

The United States and Canada require regulated local companies to offer local service at a flat rate, but most other OECD countries have separate rates for renting a telephone line and placing local calls. Local calling rates generally vary greatly by time of day. As liberalization has proceeded, most countries have "rebalanced" their rates, that is, have raised local line rates and reduced carrier access charges for long distance. In most countries, local line rates for small and medium businesses are higher than local residential access rates, but this price discrimination is being reduced as entrants begin to exploit the arbitrage opportunities created by such a regulation.

In the United States and the European Union, local residential rates have remained fairly steady since liberalization. Real local rates in the United States declined by 11 percent between 1990 and 2000, but they rose in 2001–03 owing to rate rebalancing by the FCC.[41] As a result, real local rates were about 3.7 percent higher in 2003 than in the year before local competition was launched, 1995. In Europe, average local calling rates have risen slightly since 1998 for voice calls, but the spread of zero-priced Internet calling packages undoubtedly offsets this increase.[42]

Since the United States had begun liberalizing and rebalancing long-distance rates long before 1996, its rates were much lower than in Canada or Europe in the early 1990s (see figure 9-2). However, long-distance rates fell sharply in Canada and Europe after long-distance markets were liberalized. In the United States, Canada, and Europe, long-distance rates are falling rapidly to a level that will eventually approximate local interconnection rates ("access charges") plus the long-run incremental cost of transmission and switching. Once the United States allowed the Bell companies to enter long-distance services, there was little difference among regulatory regimes across the three geographic areas, and the results have been rather similar.

The major differences are to be found in the market-opening strategies used in local telecommunications. As mentioned earlier, none of the regulatory regimes has been very successful in inducing facilities-based entry into local narrowband (voice) communications. The U.S. attempt to accelerate entry through liberal resale and unbundling policies cannot be judged a success at this juncture. Canada's more modest unbundling policy has yet to produce dividends, but it has not seen as many failed entrants as has the U.S. regime. The European policy has been the least aggressive in promoting local entry. In all three areas, entry has had little effect on local rates.

One difficulty in measuring the prices of telecommunications services across countries is that price indexes tend to be inconsistent because calling plans vary from carrier to carrier within and across countries. In some countries, the price of local calls varies with the time of day, or local calls may even be free. Subscribers may also have a choice of different mobile and fixed telephony calling plans, and these plans may differ across competitors. In addition, subscribers' calling patterns vary considerably within and across national borders.

The OECD attempts to solve these problems by calculating telecommunications price indexes for "baskets" of services for both businesses and

Figure 9-2. *Prices for National Calls in Canada, the European Union, and the United States, 1985–2001*

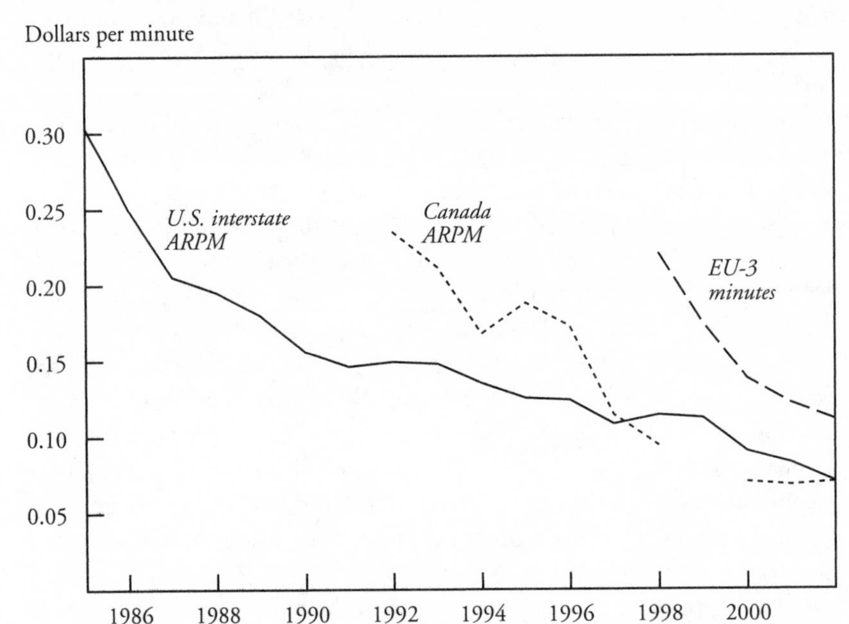

Source: United States: FCC, *Telephone Industry Revenues, 2002;* Canada: CRTC, *Annual Competition* reports; European Union: European Commission, *Annual Reports on the Implementation of the Telecommunications Regulatory Package,* Annex 1.

residences that are held constant across countries, but these indexes often use the carriers' tariff rates rather than actual transaction prices. For U.S. rates, for example, the OECD generally uses the rates of the incumbent carrier in New York (Verizon). Despite these infirmities, the OECD price indexes for 2002, as calculated for the European Commission's Directorate General for Information Society, may provide some useful insights into the effect of competition on rates (table 9-6).

One point to note is that countries with the largest entrant shares of the local and long-distance markets—Sweden, the United Kingdom, and Denmark—tended to have the lowest rates in 2002. Portugal and Greece have little domestic competition, but Greece has much lower rates than Portugal, and they are not appreciably above the EU average. Despite its long record of promoting competition, the United States has marginally higher rates than does the European Union, even though U.S. rates are based on

Table 9-6. *Residential Telephone Rates in OECD Countries*
Euros/year at PPP (purchasing power parity)

Country	*1998, old basket*	*2002, old basket*	*2002, new basket*
European Union			
Austria	494	445	646
Belgium	464	447	549
Denmark	309	303	395
Finland	323	366	490
France	387	401	522
Germany	440	346	494
Greece	523	406	675
Ireland	404	406	503
Italy	441	436	613
Luxembourg	269	332	411
Netherlands	393	379	477
Portugal	668	609	838
Spain	636	446	592
Sweden	280	291	378
United Kingdom	379	324	478
Other			
Japan	525	368	584
United States	392	395	573

Source: Teligen, prepared for European Commission (August 2002).

a state (New York) where competition is more intense than in the average state.

OECD data on broadband rates for incumbent telephone carriers, cable modem providers, and other carriers provide some further surprises, as table 9-7 shows for services offering speeds of about 1 megabit a second. (Note, however, that these data do not reflect promotional allowances or installation costs, if any, and that much lower rates are often available for services with very low speeds.) It seems that Germany has lower DSL rates than Denmark, despite the fact that Denmark has cable modem competition and Germany has very little. Most strikingly, the Japanese and Korean prices of DSL offering 8 megabits a second are lower than the prices for only 1 megabit a second in most countries. Note also the wide range in DSL prices, but the relatively narrow dispersion in cable modem prices across countries. At this early stage of broadband development, pricing the

Table 9-7. *Broadband Prices, October 2003*[a]

	DSL		Cable modem	
Country	*Download speed (kilobits/second)*	*Price (U.S. dollars/month)*	*Download speed (kilobits/second)*	*Price (U.S. dollars/month)*
European Union				
Belgium	3,000	61.58	1,024	49.18
Denmark	1,024	80.24	1,024	65.84
Finland	1,024	64.52–126.01	525	55.46
France	1,024	45.27–84.78	1,280	45.16
Germany	768	39.50–56.53	1,024	44.14
Greece	1,024	171.62–193.66	n.a.	n.a.
Ireland	512	231.45	600	45.27
Italy	1,200	73.51	n.a.	n.a.
Luxembourg	1,024	137.71	1,024	75.83
Netherlands	1,024	50.88–58.86	1,500	52.18
Austria	768	67.80	1,024	71.31
Portugal	1,024	142.47–169.51	1,024	91.61
Spain	1,024	157.56	1,024	104.91
Sweden	2,048	54.20–57.34	1,024	42.25
United Kingdom	1,024	46.79–187.25	1,024	55.76
Other				
Australia	1,500	97.59–117.12	Unspecified	45.53
Canada	1,024	26.81	1,500	34.48
Japan	8,000	21.65	n.a.	n.a.
Korea	8,000	33.25–33.61	10,000	30.16
United States	1,500	29.95–69.95	1,500	42.95

Source: OECD, *Benchmarking Broadband Prices in the OECD*, 2004 (www.oecd.org/searchResult/0,2665,en_2649_37409_1_1_1_1_37409,00.html).

a. Plans shown are for unlimited use.

service is subject to a number of tactical decisions by sellers who are attempting to promote a new service and obtain market share.

Investment in Communications Equipment

Under their less aggressive regulatory approach to liberalizing telecommunications, the European Union and Canada have experienced a more modest amount of facilities-based entry than the United States. Hence capital investment has expanded at a slower pace in these countries. This pat-

Figure 9-3. Growth in Real Spending on Communications Equipment, Selected OECD Countries, 1990–95 versus 1995–2000

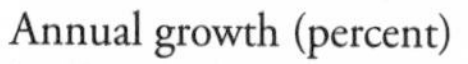

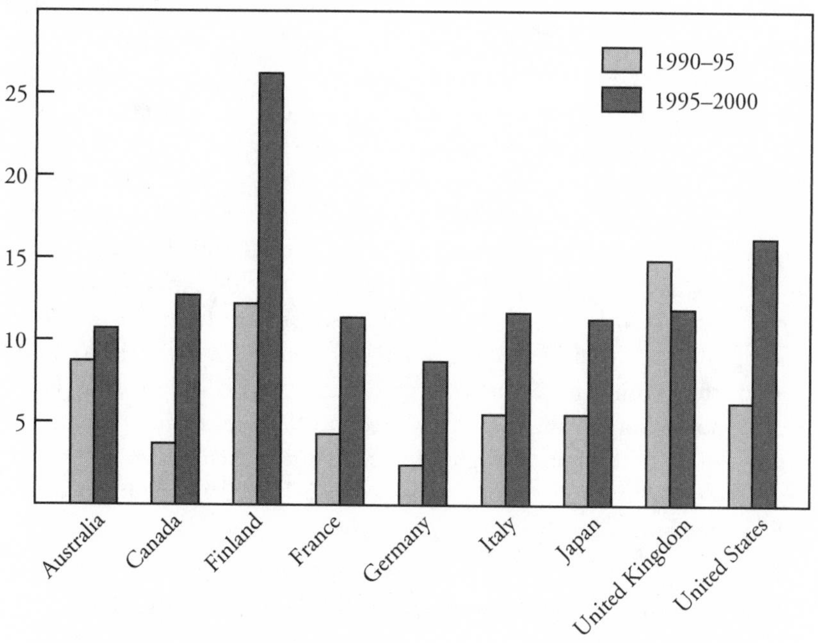

Source: Colecchia and Schreyer, *ICT Investment and Economic Growth in the 1990s* (2001).

tern is reflected in an OECD study of the growth in *real* capital spending on communications equipment, which used an estimate of "harmonized" prices for such equipment across countries.[43] The results are shown in figure 9-3. With the exception of Finland, which emerged from a deep recession in the early 1990s, all of the surveyed countries saw lower growth in communications capital spending in 1995–2000 than did the United States. The real growth rate was 50 percent greater in the United States than in Canada, France, Germany, Italy, and the United Kingdom.

Between 1995 and 2000, U.S. telecom capital expenditures grew at an annual rate of 12 percent a year in current dollars, while nominal telecom output grew by only 6.4 percent.[44] For the previous eight years, output and capital spending growth were virtually identical (recall figure 3-3). Although the growth in capital expenditures was not as great in France, Germany, Italy, and Japan, investment surged in these countries also. The subsequent collapse in investment spending after 2000 was at the heart of the 2001–03 crisis in world telecommunications.

The excessive surge in U.S. capital spending is notable. Had capital expenditures grown at a nominal rate of only 8 percent a year between 1995 and 2000, generating a real growth rate closer to that experienced in the other major OECD countries, the United States telecom sector would have spent $54 billion less on new capital equipment between 1996 and 2000. Capital spending by the largest long-distance companies rose from $10 billion in 1996 to approximately $34 billion in 2000, while CLECs spent an estimated $43 billion over the entire period. Had the long-distance companies kept their annual capital expenditures at their 1997 level of $15 billion, they would have spent $36 billion less and would be in somewhat better financial health today. Had CLECs invested only half as much in wasted collocation facilities and other investments over the 1996–2000 period, they would have spent approximately $21.5 billion less. These more modest growth rates would have reduced U.S. telecom capital expenditures by $57.5 billion, or slightly more than required to have generated U.S. capital-spending growth equal to that achieved in the other major OECD countries.

Liberalization's Effect on Incumbent Telephone Companies

When the telecommunications sector was liberalized in the 1980s and 1990s, most of its assets were in the hands of the national carriers, that is, the incumbent operators. These carriers were privatized slowly and then forced to face a more competitive marketplace. The liberalization process began in North America, but it spread quickly to New Zealand, the United Kingdom, and Japan in the 1980s, followed by the European Union in the 1990s. When Europe liberalized its markets on January 1, 1998, the worldwide telecom stock market "bubble" was already forming. Thus the incumbent carriers were forced to deal with two new but related forces: competition and a frenzied financial climate. Not surprisingly, the managements of these carriers formulated a variety of strategies to cope with this new environment.

In the European Union, the large incumbent carriers were privatized at very different rates. Telecom Italia, British Telecom (BT), and Telefonica (Spain) became fully private companies fairly quickly. On the other hand, France Telecom (FT) and Deutsche Telekom (DT) still have substantial government ownership. Despite this continued government ownership, the managements of FT and DT were among the most aggressive in pursuing mergers and acquisitions and in bidding for spectrum. These firms were most affected by the telecom bubble, as figure 9-4 shows.[45]

The two worst performers among the large foreign ILECs in terms of the value of their equities were British Telecom and NTT, the Japanese incumbent. British Telecom was very aggressive in trying to capitalize on the telecom boom of 1998–2000, but it did so in a different way than most of its EU compatriots. It invested heavily in alliances and joint ventures around the world, committing £5.8 billion (about US$8 billion) to such ventures in 1998–2000. In addition, its capital expenditures were £10 billion in these three years. These were very large sums for a carrier whose revenues were only £16 billion in 1998. NTT, on the other hand, pursued an aggressive strategy of acquiring interests in foreign wireless companies and a U.S. Internet company, Verio. In fiscal 2001 the company spent ¥1.9 trillion (about US$17 billion) on such acquisitions. In 2002, realizing that it had overpaid for many of these assets, it began to take large write-downs for them.

The value of all telecom stocks peaked in early 2000, but the values of FT and DT rose the most markedly among the large incumbent carriers in the world, increasing by about 400 percent in just over two years. The market did not take kindly to BT's and NTT's aggressive investments, however, and BT stock rose a modest 100 percent during this period, while NTT's stock posted similarly modest growth.

By contrast, Telefonica and Telecom Italia were not very aggressive in acquiring wireless or other assets after liberalization. As a result, their equities rose less during the 1998–2000 frenzy. When the bubble burst, Telefonica's equity price fell, but it was still substantially above its December 1997 value in mid-2004. Telecom Italia's equity price showed similar stability. On the other hand, the equity values of BT, NTT, FT, and DT fell more sharply after February 2000, when it became obvious that their various investments in joint ventures, wireless assets, and new 3G spectrum had been purchased at prices that were far too high. In the last two years, all of these companies have been writing down these assets, selling a variety of other assets, and aggressively trying to reduce the debt they issued to make acquisitions in the past few years.

Conclusion

The world's telecom crisis in 2001–03 was driven in no small part by the major national telecommunications carriers investing heavily outside their

Figure 9-4. *Stock Performance of Foreign ILECs*[a]

Index, December 1997 = 100

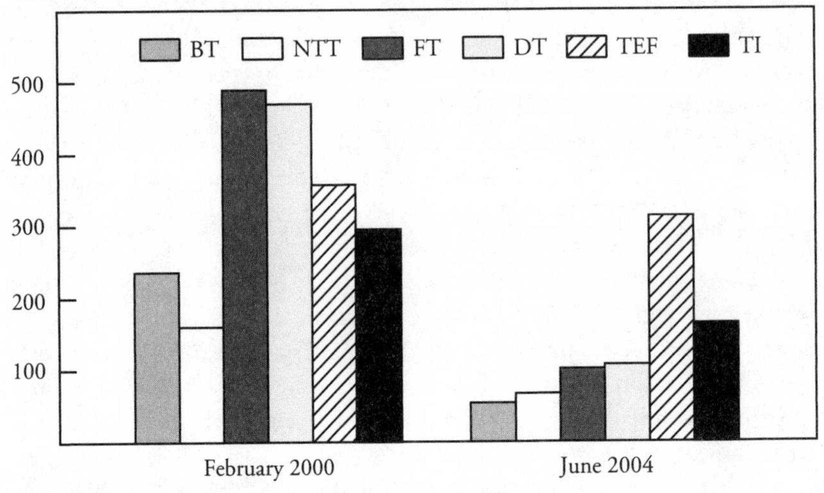

Source: www.finance.yahoo.com

a. TI quotes are for February 2000 and June 2003. BT = British Telecom, NTT = Nippon Telephone & Telegraph Corporation, FT = France Telecom, DT = Deutsche Telekom, TEF = Telifonica (Spain), and TI - TelecomItalia.

own borders in wireless and Internet-driven ventures. The incumbents that "stayed home" and attended to their traditional businesses, such as Telefonica, Telecom Italia, and the U.S. Bell companies, have survived in much better condition—although exposed to the declining revenues from traditional telecom services.

None of the OECD regulatory policies examined in this chapter favored entrants as much as those erected by the United States. As a result, they have fostered less narrowband competition than in the United States, but they also left fewer bankrupt entrants littering the telecom landscape in 2003–04. In most countries, mass-market competition in voice services comes from wireless services, not new wire-based entrants.

Only Japan has used local-loop unbundling as aggressively as the United States, but Japanese policy has been confined largely to broadband services. Given the enormous losses being recorded by the parent of the principal Japanese broadband entrant, Yahoo-Broadband, it is far from clear that this policy will achieve its ultimate goal. Korea has the greatest broadband penetration of any country in the world and boasts three major broadband competitors despite having no unbundling policy until recently. The EU unbundling policy is having limited success in developing broadband competition.

The world telecommunications sector is now beginning to recover from the excessive optimism created by the Internet in the late 1990s. The overextended national carriers, such as France Telecom and Deutsche Telekom, have retrenched considerably. Global Crossing has essentially disappeared as a global company, and most of its international assets have been acquired in bankruptcy by Singapore Telephone. Most of the new local entrants in Europe and the United States have been reorganized or have simply disappeared. Wireless carriers continue to develop new services, although many outside the United States are still bruised from paying enormous prices for 3G spectrum that they are only now beginning to use. And in countries like the United States and Canada, where cable television connects two-thirds of all households, the cable operators are now major players in broadband Internet communications and are preparing to offer voice services as well. The crisis is essentially over for many of these players, but its after effects linger on in the form of depressed investment spending.

10 A Look Back and a Look Forward

No one looking back over the past nine years of meteoric events in the telecom sector can confidently predict what will happen in the next nine years. Who can say how the politics of regulation will affect the sector after the rather disappointing impact of the 1996 Telecom Act on traditional wire-based telephone operations? In hindsight, perhaps the stock market bubble and excesses of 1998–2000 could not have been avoided, for the Internet created incredible expectations. Regulators may not have created it, but they did contribute to the bubble by encouraging entrants having little hope of survival.

Now, five years after the bubble burst, some important lessons are emerging from the shakeout that ensued. Clearly, it is not productive to try to create competition by allowing new, relatively small firms to exploit regulatory arbitrage opportunities in delivering yesterday's services. The emphasis should be on facilitating investment in new, high-speed services and the applications or content that may be required to amortize such investment.

Regulated Competition Has Not Worked

As the preceding chapters have demonstrated, the U.S. experience with telecom liberalization since 1996 has not been a happy one. For nine years, regulators have attempted to create competition in local services and have succeeded mostly in sowing confusion and uncertainty in a sector in the grip of declining revenues. Because of the new regulatory provisions in the 1996 Telecommunications Act, a legal and political battle developed among the incumbent local telephone carriers—principally the Bell companies—and the long-distance companies and the new local entrants (the CLECs). The latter two groups are now in steep decline after spending more than $200 billion on capital facilities between 1996 and 2003.

In the post-1996 era of telecom regulation, considerable effort was put into creating an environment conducive to the entry of new carriers into fixed-wire local markets.[1] The entrants this attracted offered little in the way of innovation or new services. They were mainly interested in exploiting the arbitrage opportunities created through the regulation of wholesale and retail rates, and most of them failed with a vengeance when the telecom stock market bubble burst in 2000–02. The regulators' response was to try to protect some of their progeny by creating larger potential arbitrage margins through the unbundled network element (UNE) platform. Unfortunately, this strategy simply allowed the arbitrageur entrants (that is, those with no local capital facilities) to grow at the expense of those who had already invested billions of dollars. Regulation thus helped to condemn to an early death the only set of new entrants who could possibly innovate through their own network facilities, though most of the latter firms would no doubt have failed eventually because they could not differentiate themselves from the incumbents. Now the courts have condemned the UNE platform as well, eliminating even the lure of regulatory arbitrage.

As it turned out, the regulatory strategy of the Federal Communications Commission (FCC) and of the state regulators to promote entry into local fixed-wire services was not only wasteful but unnecessary. Meaningful mass-market competition did not develop through unbundled network elements or their platform. For the most part, these policies simply transferred billions of dollars from incumbent telephone companies to fund marketing campaigns required to sell the same service under a different name. Instead, competition has developed in ways totally unanticipated by regulators, namely through unregulated wireless providers and cable broadband platforms. Fortunately, federal and state regulators were denied the right to regulate wireless rates and services, and cable television regulation was scaled back considerably in 1996. As a result, the wireless sector began to launch an unregulated competitive rate war and to offer national calling plans as soon as it could consolidate into six national players after new frequencies were auctioned in 1995–96. At about the same time, cable operators began to expand their capabilities to meet satellite competition and were thus poised to be early movers in the broadband race. The result has been a dramatic shift of telecom revenues from traditional long-distance carriers and, to a lesser extent, incumbent local companies to wireless carriers and cable operators.

When the regulators could no longer keep the Bell companies out of long distance after seven years of Section 271 proceedings under the 1996 act, the Bell companies added further to the assault on long-distance company revenues. And when unregulated voice over Internet protocol (VoIP) services began to appear, cable television operators were forced to offer voice services rather than allow third-party carriers to siphon revenues from their cable modem customers. None of this new competition required the guiding hand or patient nurturing of regulators.

Convergence or Divergence?

The popular literature on telecommunications is replete with observations on the trend toward "convergence" of voice, data, and video services. Such convergence has clearly not occurred yet. More than 90 percent of U.S. households still use a telephone connected by copper wires for some of their voice calls, but only about 10 percent of these households use the same copper wires for broadband Internet connections. About 65 percent of Americans watch television over a cable television system; and another 21 percent subscribe to satellite television.[2] Since digital subscriber line (DSL) speeds are generally 1 megabit per second or less, no one can yet obtain much video entertainment over the traditional copper-wire telephone network. Although nearly one-half of Americans have a cell phone, which they use for voice calls and some enhanced services, current 2G speeds are not fully competitive with the fixed-wire broadband Internet connection. Nor are they suited for receiving video signals even as U.S. carriers begin to deploy 3G technologies.

The limited degree of convergence in the United States thus far is found on cable television systems. Cable operators routinely offer distributive video services in various packages (including pay-per-view) and broadband cable modem services. All of the major system operators are beginning to deploy voice services through VoIP. Although cable systems had only about 3 million telephone (voice) subscribers at the end of 2003, this number will surely increase in the next year or two, prodded by the entry of independent VoIP suppliers, such as Vonage and Skype. However, no U.S.

cable operator can currently offer the most flexible and growing form of voice service: cellular telephone service. In short, there are still practical limits to convergence.

Traditional telephone companies might be able to induce convergence over their networks by deploying fiber to or near the home, which would then make them major competitors of the cable and satellite carriers for distributive video services. They would also be able to deliver services now downloaded from the Internet at much higher speeds. Indeed, with high-speed fiber connections, traditional video services might gravitate to the Internet or some similar interconnected network.[3] If this trend develops, the cable companies are also likely to extend fiber directly to their subscribers.

Of course, satellite companies such as DirecTV or DISH Network may develop their own high-speed Internet services through the use of spot beams, but they could not penetrate the voice market or even a variety of high-speed two-way services from their geostationary satellites. Given the commercial failure of several low-orbiting satellite systems such as Globalstar and Iridium, voice services are unlikely to shift to satellite delivery, but even if they did, there would be no convergence with the video platform that resides in geostationary orbit.

The three major platform operators may have little to fear in the way of competition from terrestrial fixed wireless systems. Although much has been said about the possible deployment of "mesh networks" using unlicensed spectrum for delivering signals over short distances, there is still no widespread commercial development of such systems. One must conclude that for the foreseeable future, cable operators, telephone companies, and cellular wireless carriers will compete actively for different aggregations of telecom services, but that no one platform will likely be the beneficiary of convergence to the extent that it leaves the others at a severe disadvantage.

While the technological joint economies across all services may not be large enough to induce the convergence of all major communications services, the joint economies in marketing are likely to doom entrants who offer only a limited array of services. Subscriber switching costs for existing services are substantial, and customers must be persuaded by expensive marketing campaigns to abandon their current service for a new offering that is often similar if not identical to the service they currently have. The obvious solution is to spread these marketing costs across several services by offering the potential subscriber a bundle of services. Without such a bundled offering, many CLECs have simply been unable to cover their start-up costs, which include very large marketing outlays.

Thus single-service companies are likely to have an increasingly difficult time competing with the incumbent telephone companies and cable companies. As the Bell companies move to expand their packages with their own wireless services, the single-service voice carriers will find it even more difficult to amortize their operating costs.

Competition's Effect on Revenue Growth

Despite widespread optimistic views of the likely effects of the Internet on telecommunications, the past eight years have witnessed remarkably little growth in telecom revenues. Since 1996 the wireless carriers and cable television companies, the two largely unregulated groups of carriers, have experienced annual revenue growth of 19 percent and 9 percent, respectively (see table 10-1). By contrast, wireline telephone carriers have registered a slight decline in revenues, which is now accelerating. The reason for this disparity in growth is aggressive price competition among all carriers; it is driving rates down and causing a shift from wired telephony to wireless voice services, which means wire-based telephone companies are losing access lines. Wireless and cable companies, on the other hand, continue to attract new subscribers and to offer new services, thereby generating revenue growth.[4]

Competition in traditional narrowband (voice/data) services necessarily depresses carrier revenues because the demand for such services is price inelastic. With these traditional communications services so widely dispersed throughout U.S. households and businesses, growth must come from the development of new services. As cable develops its broadband and voice over Internet services, in addition to more and more channels of distributive video, its revenues per subscriber will continue to grow. Similarly, as wireless offers new features and increasingly serves the long-distance voice market, its revenues per subscriber will surge beyond even the levels of the past three years.

Note the steady increase in cable revenues in recent years in response to the addition of video channels and, in particular, cable's success in selling cable modem services. Cable operators can also sell advertising in their program offerings. Wireless carriers, too, have enjoyed increasing revenues per subscriber since they began offering long-distance service through nationwide pricing plans. However, this growth is likely to abate as the Bell companies expand into long-distance service in the wake of the Section 271 approval process. Wireless carriers will clearly need new, advanced services to allow their revenues per subscriber to continue to grow, but they still have several years of revenue growth available from attracting a large share of the Americans who still do not have a cell phone.

The most endangered species in the communications sector is the wire-based company that tries to survive by offering a single service or a limited set of services, such as local voice-data connections or long-distance service. The revenues available from wire-based services, including DSL, have been declining rapidly for several years.[5] Thus the revenues available from telecom services for all wire-based carriers, including the incumbent telephone companies, the CLECs, and the long-distance companies, are declining, and they could decline further as severe price competition continues. The only survivors among these companies are likely to be those that can offer multiple services, including wireless and broadband Internet connections. Ultimately, they may have to build fiber very close to or at the subscriber's location to be able to compete with the multiservice cable companies.

The Telecom Sector Is Quite Healthy

The financial press is still replete with references to the "troubled" telecom sector suggesting that the entire sector collapsed after the bubble burst in 2000 and never recovered. Although such reports may assuage financial market participants who bid up the stock prices of Global Crossing, WorldCom, the various new local entrants, and the telecom equipment manufacturers for little apparent reason, they simply do not fit the facts of the current U.S. telecommunications sector. Competition and a harsh dose of reality have devastated three of its subsectors: the CLECs, the long-distance companies, and the once-large equipment manufacturers. The CLECs and long-distance companies will not recover from these twin assaults because they simply cannot compete with the cable companies, the wireless carriers, and the large incumbent local carriers without their own local networks and a much broader product line.

The two large North American equipment companies are also in dire condition because they extended substantial credit to many of these failing carriers, and they have not been able to adapt to the changes in technology that are driving packet-switched broadband services. A simple chart (figure 10-1) demonstrates how the fortunes of equipment producers rose and fell with the telecom bubble that first supported and then destroyed the CLECs and the long-distance carriers. The total market value of Nortel

Table 10-1. *Revenue Growth in Three Communications Sectors, 1996–2003*[a]

Sector	1996	1997	1998	1999	2000	2001	2002	2003	Average annual growth (percent) 1996–03	2000–03
Wire-based telecommunications carriers										
End-user revenues	151.5	158.3	167.0	171.0	171.7	167.6	157.4	149.2	–0.22	–4.68
Revenues/line	931	926	934	919	899	872	826	801	–2.15	–3.85
Wireless carriers										
Revenues	23.6	27.5	33.1	40.0	52.5	65.3	76.5	87.6	18.74	17.07
Revenues/subscriber	608	553	532	516	537	529	569	592	–0.38	3.25
Cable television companies										
Revenues	27.7	30.5	33.5	36.9	40.9	43.5	49.4	51.3	8.80	7.55
Revenues/subscriber	428	463	500	539	590	596	672	698	6.99	5.60

Sources: Wire-based telecommunications carriers: FCC, *Telecommunications Industry Revenues,* annual editions; wireless telecommunications carriers: CTIA, *Semi-Annual Wireless Industry Survey,* downloaded from www.wow-com.com/industry/stats/surveys/; cable television: NCTA, *Industry Overview*, downloaded from www.ncta.com/docs/pagecontent.cfm?pageID=86.

a. Revenues in billions of dollars/year; revenue per subscriber or per line in dollars/year.

Figure 10-1. *Market Value of the Two Major North American Telecom Equipment Suppliers, Selected Years*

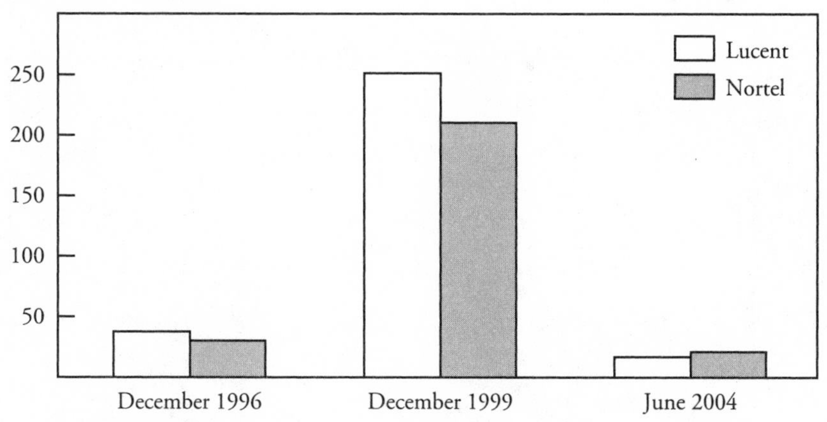

Source: Yahoo finance.

and Lucent soared during the "bubble," but by the end of 2003, their value had fallen to less than two-thirds of their value at the end of 1996.[6] The stock market value of these two erstwhile titans has declined by more than one-half since the end of 1996.

By contrast, the cable companies, wireless carriers, and incumbent local telephone carriers remained quite healthy even after the bubble burst. As figure 10-2 shows, their market values per subscriber or subscriber line changed remarkably little over the 1996–2003 period. The cable companies' value per subscriber has risen sharply since 1996, while the large incumbent local (Bell) carriers' value per line has risen only marginally. Only the wireless carriers, among the obvious survivors, have suffered a reduction in value per subscriber, but they have enjoyed phenomenal growth in subscribers. Thus their total market value has nearly trebled. In each case, however, the survivors' market value per line or subscriber appears to be at least as great as the reproduction cost of their assets, hardly a condition typical of a sector in "crisis." The actual crisis occurred in those subsectors created mainly by regulation, nurtured by regulators, and propelled by speculators. The wireless, cable, and integrated telephone carriers may have been swept along in some of the exuberance, but they never tumbled into the subsequent abyss.

Figure 10-2. *Market Value per Subscriber or Line for the three Groups of Survivors, 1996 and 2003*

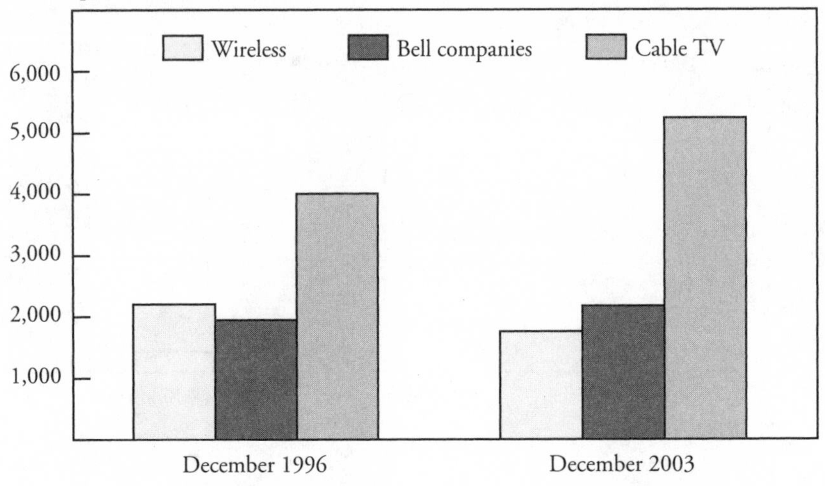

Source: Yahoo finance.

From Regulation to Protection?

The architecture erected by U.S. telecom regulators is now crumbling. They have no strategy for continuing to prop up failing local entrants since the courts ruled that the 1996 act does not entitle entrants to access to the entire incumbent network at large wholesale discounts. The FCC has been unable to fashion a broadband policy that applies symmetrically to cable modem services and incumbent telephone companies' DSL services. And the FCC and the states are scrambling to deal with the obvious threat of VoIP, which can be offered to U.S. broadband subscribers by carriers located in any country in the world, perhaps without all of the taxes and surcharges placed on traditional telephone service in the name of "universal service."[7]

Students of regulation are generally wary of regulated competition. Airline, trucking, railroad, and petroleum regulation in earlier decades became a form of cartel management, keeping prices artificially high and entry low in order to protect competitors.[8] In all likelihood, U.S. telecom regulators will soon be faced with demands to erect an umbrella over prices, much as the International Commerce Commission and Civil Aeronautics Board did for transportation companies in the twentieth century and as the Canadian Radio-Television and Telecommunications Commission is already doing for Canada's weak local telecom entrants. As the Bell companies increasingly market bundles of services, including local, long-distance, wireless, and DSL services, the faltering long-distance companies and CLECs will begin to argue that such competition is "predatory," or at least "unfair." When the cable companies offer similar packages of video, voice, and broadband services, pressure will build to regulate them too, to protect the few remaining struggling new local entrants and older long-distance companies.

The United States has a long and undistinguished history of using politics to protect industries confronted by new competition. Trade protection for textiles and steel are perhaps the best-known modern examples. As of October 2004, a large number of U.S. goods producers are lobbying to stop "jobs" from moving to China. In the 1930s, the Robinson-Patman Act was passed to protect small grocery stores from the more efficient chains that now provide the public with low-cost food distribution.[9] Television broadcasters successfully lobbied the FCC to prevent new, higher-frequency bands from being utilized to increase video competition and, subsequently, to protect them from cable television from 1972 through 1989 by limiting what cable systems could offer.[10]

In each of these cases, protection occurred because politicians were forced to respond to public sympathy for the established companies and their employees. In the current U.S. telecom sector, the long-distance companies and new local entrants operate across the country and employ thousands of workers, many of whom sit at telemarketing terminals desperately trying to prevent the flight of customers to the integrated facilities-based carriers. Politicians may well respond to the plight of these entities and their employees by asking regulators to protect them from "unfair" competition from the giant cable companies, Bell companies, and wireless carriers. If this fails, new legislation could be drafted—repeating the history of the Robinson Patman Act in 1936—in the telecom post-2000 "depression."

Even the Bell companies are far from secure. If the new VoIP services over cable induce subscribers to drop their traditional local service, the Bell companies may see their revenues continue to decline unless they can retain customers through an aggressive broadband strategy. If they fail in this attempt, they too could become advocates of protective regulation, arguing, for example, that the new cable company telephone services compete "unfairly" because the cable companies have not had to suffer the burdens of being the carrier of "last resort" or to support the failing CLECs with below-cost wholesale services. The dynamic technological change in this sector could leave many companies exposed to the cruel winds of change from which they would be induced to seek government protection.

The economic lesson from the history of regulation is that regulation and competition are a bad emulsion. Once the conditions for competition exist, it is best for regulators to abandon the field altogether. This is particularly true in a sector that is undergoing rapid technological change and therefore requires new entry and new capital. The politics of regulation favor maintaining the status quo, not triggering creative destruction.

U.S. Telecom Regulators as Tax Collectors

The regulation of prices and entry conditions generally presumes the threat or existence of a "natural monopoly" or a "dominant" market position. But in the current telecom marketplace, there is so much competition across platforms—fixed-wire, wireless, and cable television—that carriers are not likely to be able to exercise monopoly power in traditional voice/data services or in the new broadband services. Indeed, it is not clear that regulators are constraining prices in any of these markets. The pressure for regulation today derives from a political demand to keep rates *high*, not to restrain monopoly power.

Among the most important issues facing U.S. telecom regulators today are those that involve the levying of taxes on telecom services to support a variety of "universal service" objectives. These taxes are of two forms: (1) explicit levies on interstate and international revenues and (2) implicit subsidies through regulatory rate distortions.

The 1996 Telecommunications Act instructed the FCC to move toward explicit, transparent subsidies to support low-income subscribers, high-cost rural telephone companies, schools, libraries, and rural health facilities. The FCC complied by establishing a universal service levy on all telephone revenues, but the federal courts subsequently ruled that the tax base for such a program must be limited to interstate and international revenues. The current tax rate for these programs is now about 9 percent of these revenues, which generates about $6 billion in annual receipts to sprinkle among the politically favored constituent groups. The schools and libraries allocation is absorbing about $1.7 billion a year of these funds, and the rural, high-cost telephone companies are receiving $3.3 billion a year.[11] As the revenue base declines because of competition in interstate and international services, the tax rate must be increased or the disbursements must be reduced. Obviously, the development of free long-distance services on wireless systems in off-peak hours and VoIP are major threats to these programs.

A second form of subsidy generated by the current regulatory system is found in the regulated rate structure. Regulators have traditionally kept long-distance rates and some business rates high to cross-subsidize high-cost rural telephone services. The most obvious of these implicit subsidies is now found in the switched access rates charged by local carriers to originate and terminate long-distance calls. These rates are often as high as 3 cents a minute at each end of an intrastate call and about 1 cent a minute for interstate calls originating and terminating on nonrural local carriers' networks. The marginal cost of switched access is undoubtedly less than 0.1 cents a minute for today's electronic switching systems. In 1996 the large incumbent local carriers reported $20 billion in switched access revenues. By 2002, the most recent year for which data are available, they had declined to $10.5 billion in switched access charge revenues.[12] Today they are still about $8 billion.

These elevated access charges on fixed-wire networks are in severe danger because of the enormous shift of long-distance services to wireless carriers. By originating a long-distance call on a wireless handset, the subscriber reduces the number of access minutes by at least 50 percent. If the subscriber calls another wireless subscriber, the reduction is 100 percent. Not only are switched access minutes not growing as rapidly as in the 1990s, but they are declining steadily, and—as already shown—access revenues are following the same path. In response, the fixed-wire carriers have attempted to engineer a regulatory solution known as "bill and keep" that would reduce access charges to zero.[13] Each local carrier would pay to transport its traffic to or from a node on the network, and the loss in access revenues would be recovered from a fixed charge levied on the local subscriber. Unfortunately, this proposal is now faltering, in part because the local carriers do not believe they can pass on the increased fixed charge to their customers.[14] This surely suggests that the incumbent fixed-wire "monopolists" have very little market power in today's environment of wireless and VoIP competition. As a result, the FCC and state regulators are likely to have difficulty persuading incumbents to reduce access charges appreciably; they will simply have to watch as their second source of subsidies withers away in the face of wireless and VoIP competition.

The most important regulatory battles of the next few years will revolve around attempts to preserve these subsidy programs in the face of advancing competition. One approach would be to tax VoIP and other competitive services in the same manner that fixed-wire and wireless revenues are taxed to fund the universal service program. Another might be to regulate VoIP or wireless rates so as to prevent unfair, "predatory" pricing of long-distance voice services, that is, to keep rates high much as the Interstate Commerce Commission and the Civil Aeronautics Board did for decades before their franchises were ended in the 1980s.

An Opportunity to Deregulate?

Federal and state regulators may find this a propitious time to consider deregulating telecommunications services. The traditional policies of keeping business and long-distance rates high to support below-cost residential access to the network are failing as long-distance rates plunge owing to wireless competition and the emerging competition from VoIP. Eight years of competition from the new CLECs has had little effect on prices and therefore on these policies. However, the new competition from wireless and VoIP, if it succeeds in driving telephone rates toward costs, will undermine the regulators' attempts to transfer wealth through the rate structure in the name of "universal service." Such a rebalancing of rates would be enormously beneficial in terms of economic welfare, generating as much as $7 billion in static efficiency gains to the economy.[15]

Equally important, as telecom revenues stagnate, the politically driven demand for *explicit* universal service subsidies can only be satisfied by higher and higher "contribution rates" (that is, taxes) levied on interstate and international revenues that appear monthly on subscriber bills. These contribution rates are already about 9 percent of revenues and generate more than $6 billion in annual subsidies for a variety of purposes.[16] Given the relatively high price elasticity of demand for interstate and international services, the cost of these taxes is extremely high, costing the economy between $3 billion and $7 billion a year over and above the $6 billion raised by the program.[17] It is difficult to see how these extraordinary costs can be justified by regulators who are sensitive to ratepayers' concerns about increases in telephone bills.

Moving away from the implicit and explicit universal service policies could therefore generate as much as $14 billion in economic efficiency gains. Even if deregulation of all voice services created some opportunity for telephone companies to exert market power, it is difficult to see how any such monopoly pricing could cost the economy $14 billion a year. Assuming that the demand for local service has a price elasticity of demand of –0.03 and a current price of $25 a month, even a 50 percent increase in local rates for every household in the country would create a deadweight loss of only about $100 million a year. Given the rise of wireless and VoIP, the price elasticity of demand for local service is surely increasing, but it may increase so much that incumbents will have little ability to raise local rates anyway.

The new problems posed by VoIP telephony and cable television delivery of high-speed services have already ensnared the FCC in a difficult set of issues. How can it continue to regulate DSL and circuit-switched voice telephony and not regulate VoIP and cable modem services? If it forbears from regulating the latter services, what will happen to its contribution base for universal service charges? Will states have to allow ILECs to raise local rates to residential consumers in order to offset the loss in access charges and long-distance revenues that have been supporting residential rates for decades?

The obvious direction for regulators is a phase-down of all telecom regulation. The unbundling regime could be phased out in three years, as some observers have suggested.[18] All carrier compensation could be moved in the direction of bill-and-keep as VoIP develops, and then all regulation of voice services could be eliminated.

The Future: Broadband, Content, and Vertical Integration

The telecommunications sector is clearly undergoing a major transformation from narrowband voice/data service to broadband service. As the price of all voice services falls dramatically to marginal cost, they become a low-cost complement to much more complex services delivered over higher-speed networks. Eventually, all information will be transformed into packets for delivery over packet-switched networks. The architecture of these networks is still evolving as cable television and telephone companies adjust to the new market realities and as wireless options such as WiFi proliferate. These new broadband networks require large capital expenditures on fiber optics, new switching architectures, and more complex equipment on the customer's premises. Companies can no longer expect the investment environment to be regulated as it was in the twentieth century, however. New players will certainly emerge with new technologies, but without an assured regulated rate of return. The risk of investing in network assets will rise substantially.

If the new broadband platforms are to be built and extended to a large fraction of the population, the platform owners must be persuaded that there is sufficient demand for the services they offer. This, in turn, requires that new applications ("content") be developed to drive this demand. As discussed in chapter 8, innovations in other new technologies with these "network effects" in the twentieth century were generally accompanied by vertical integration between the owners of content (or the upstream input) and the owners of the distribution system.[19] In most of these cases, the developers of the new technologies were innovative entrepreneurs, such as the automobile industry's Henry Ford or cable television's John Malone.

Economists now recognize that network effects are exceedingly important in the development of the new broadband technologies and that those who deploy the platforms should have the property rights to ration access to their platforms.[20] But will the major platform owners be successful in developing content or applications for their new networks? The major contenders for this role in the United States and elsewhere are the incumbent telephone companies and the cable television companies, neither of which have shown much imagination in these pursuits in recent years. The telephone companies have never been very entrepreneurial, nor do they have expertise in developing and marketing innovative consumer products. The cable companies were once in the vanguard of developing entertainment products for their distributive video services, but the recent stumbles of AOL-Time Warner (now "Time Warner") do not bode well for the successful integration of distribution and content in the new broadband world.

The advocates of end-to-end Internet connectivity fear that the network operators will indeed begin to exploit the network effects available from deploying new services. As discussed in chapter 8, they would use regulation to require these network companies to use nonproprietary protocols and to provide open access to their networks. Such regulation would surely reduce the incentive to invest billions of dollars in new network facilities. But even if the platform owners were to retain the right to exploit these network effects, it is still not clear that they can do so.

The difference between the current broadband revolution and earlier transformations of consumer services or products industries is that the broadband distribution platforms are being built by large, still-regulated companies—or erstwhile regulated companies, in the case of cable television. These companies are not the obvious repositories of innovation and entrepreneurial energy found in the companies run by Henry Ford, David Sarnoff, or Samuel Goldwyn in the twentieth century. There is surely little role for public policy to find the best organizational structure to exploit the relevant network effects. The markets will have to provide this service, hopefully unencumbered by regulation.

Changes of ownership may be required. For instance, Rupert Murdoch recently acquired the leading U.S. satellite broadcaster, DirecTV, from General Motors. Telephone companies could be transformed through acquisition, much as Telecom Italia was transformed when it was acquired by Olivetti. Time Warner, though not successful in combining with AOL, has media, programming, and cable television operations that could be employed in the exploitation of broadband. The capital markets are surely the best instruments for matching opportunities with the requisite management skills and entrepreneurial energies.

Appendix: Estimates of Bell Company Cumulative Capital Expenditures across States, 1996–2003

This appendix provides a summary of the regression analysis of cumulative capital investment by Bell companies across states in 1996–2003. These regressions attempt to replicate the empirical methodology of Robert Willig and his colleagues described in chapter 5, using the same variables where possible.

Data

A complete list of the variables used appears in table A-1. Willig and associates use a measure of the rate for leasing the UNE-P provided to them by AT&T. I use four measures available from public sources in addition to one from Kevin Hassett and his associates that is apparently from AT&T. I also insert separate dummy variables for each Bell company other than Qwest (U S West) in each of the regression equations. The complete results are available from the author.

Results

I attempt to replicate the basic equation of Willig and associates:

Table A-1. *Variables Used in Capital-Spending Regression Analysis, 1996–2003*

Variable	*Definition*	*Source of data*
CapEx/Pop	Change in net plant over the period divided by the state's population at the beginning of the period	FCC ARMIS; Census Bureau
Plant/Pop	Previous period's plant in service divided by the state's population	FCC ARMIS; Census Bureau
FIREA	Share of state employment in financial, insurance, and real estate sectors	Bureau of Economic Analysis, U.S. Department of Commerce
PopGrowth	Population growth rate in the state over the relevant period	Census Bureau
Unemp	Average unemployment rate over the period	Bureau of Labor Statistics
BusRev	Average revenue per business line	Gregg; Burns and Kovacs
ResRev	Average revenue per residential line	Gregg; Burns and Kovacs
Cost	Average TELRIC cost per line	FCC HCPM model
UNE-P	Monthly lease rate for UNE-P	Gregg; Burns and Kovacs; Hassett and associates
CLECShare	CLECs' share of access lines in the state as of December 31, 2002	FCC *Local Competition Report*
Reg	Five dummy variables reflecting state regulation of Bell company: price cap, price cap with freeze, rate of return, freeze with nonindexed cap, and deregulation	National Regulatory Research Institute
BLS	Dummy variable for Bell South	
SBC	Dummy variable for SBC	
VZ	Dummy variable for Verizon	
USW	Dummy variable for U S West (Qwest)	

$$\text{CapEx/Pop} = a_1 + a_2\,\text{Plant/Pop} + a_3\,\text{FIREA} + a_4\,\text{PopGrowth} + a_5\,\text{Unemp} + a_6\,\text{BusRev} + a_7\,\text{ResRev} + a_8\,\text{Cost} + a_9\,\text{UNE-P} + \Sigma\, a_i\,\text{Reg}_i + u,$$

where u is a random disturbance term and the other variables are as defined in table A-1.[1] The mean and standard deviation of each variable are shown in table A-2. The estimates of interest are the coefficients of

Table A-2. *Mean and Standard Deviation of Variables Used in Regression Analysis, 1996–2003*

	Mean	*Standard deviation*
CapEx/Pop	0.065	0.126
Plant/Pop	0.917	0.327
FIREA	0.055	0.014
PopGrowth	0.082	0.065
Unemp	4.443	0.961
BusRev	41.661	13.194
ResRev	21.727	3.768
Cost	14.046	0.758
UNE-P (Kovacs)	24.885	6.608
UNE-P (Gregg-01, Zone1)	18.427	4.830
UNE-P (Gregg-01, Zone 2)	23.028	6.491
UNE-P (Gregg-02)	21.056	4.665
UNE-P (Gregg-03)	18.088	4.091
UNE-P (AT&T)	22.691	6.021
CLEC Share	12.667	4.754
PriceCap	0.449	0.503
PriceCap-Freeze	0.224	0.422
RORReg	0.245	0.434
RateFreeze	0.061	0.242
BellSouth	0.184	0.361
SBC	0.265	0.446
Verizon	0.265	0.446

UNE-P, a_9. These are reproduced in table A-3 for the various measures of UNE-P.[2]

The results in table A-3 are shown for several measures of the UNE-P rate across states. Despite the fact that this is a regulated price, no one can agree on what the price is at any point in time. I estimate these regressions for one measure published by Anna Maria Kovacs and Kristin Burns, four published by Billy Jack Gregg, and one by Hassett and his colleagues.[3] The last study was funded by AT&T and therefore undoubtedly contains UNE-P data from AT&T, the same source used by Willig and associates.

Note that the significant, negative coefficient on the UNE-P rate variable for some of the regressions involving 1996–99 capital expenditures becomes insignificant for 2000–03 capital spending when the Kovacs and Burns

Table A-3. *Coefficient Estimates from Capital Spending Regressions, 1996–2003*[a]

Variable	*1996–99*	*2000–03*	*1996–2003*
UNE-P (Kovacs and Burns, 1,200 minutes of usage/month)			
No dummy variables	–0.0027**	–0.0018*	–0.0044**
	(–2.99)	(–1.77)	(–2.68)
With regulatory dummies	–0.0017**	–0.0015	–0.0033
	(–2.16)	(–1.44)	(–2.10)**
With regulatory and Bell dummies	–0.0017	–0.0014	–0.0030
	(–1.68)	(–1.06)	(–1.55)
UNE-P (Gregg, 2001 rates, zone 1)			
No dummy variables	–0.0043**	–0.0026**	–0.0068**
	(–2.76)	(–2.11)	(–2.58)
With regulatory dummies	–0.0038**	–0.0026**	–0.0061**
	(–2.42)	(–2.07)	(–2.52)
With regulatory and Bell dummies	–0.0037**	–0.0023*	–0.0058**
	(–2.20)	(–1.75)	(–2.24)
UNE-P (Gregg, 2001 rates, zone 2)			
No dummy variables	–0.0027**	–0.0030**	–0.0058**
	(–2.27)	(–3.51)	(–3.37)
With regulatory dummies	–0.0028**	–0.0027**	–0.0056**
	(–2.48)	(–3.04)	(–3.20)
With regulatory and Bell dummies	–0.0029**	–0.0028**	–0.0057**
	(–2.31)	(–2.86)	(–3.03)
UNE-P (Gregg, 2002 rates)			
No dummy variables	–0.0059**	–0.0024*	–0.0082**
	(–3.78)	(–1.74)	(–3.25)
With regulatory dummies	–0.0052**	–0.0026*	–0.0077**
	(–3.20)	(–1.85)	(–2.95)
With regulatory and Bell dummies	–0.0058**	–0.0027	–0.0085**
	(–3.12)	(–1.63)	(–2.83)

(continued)

UNE-P measure is used. The same is true for the rates published by Gregg in 2002. Gregg's data yield increasingly significant negative coefficients for the effect of UNE-P on 1996–2003 capital spending by the Bell companies as the date of the UNE-P price advances from 2001 to 2003. One can hardly interpret this result as showing that capital spending in 1996–99 was driven by UNE-P rates as states lowered them in *2002 and 2003*. Moreover, the coefficients on Gregg's measures of the UNE-P rate are always greater in absolute value for 1996–99 than for 2000–03 capital spending. Why would 1996–99 capital spending be more sensitive than 2000–03 capital spending

Table A-3. *Coefficient Estimates from Capital Spending Regressions, 1996–2003*[a] *(Continued)*

Variable	*1996–99*	*2000–03*	*1996–2003*
UNE-P (Gregg, 2003 rates)			
No dummy variables	−0.0090**	−0.0059**	−0.0149**
	(−4.93)	(−4.07)	(−5.72)
With regulatory dummies	−0.0091**	−0.0069**	−0.0161**
	(−4.95)	(−4.69)	(−6.16)
With regulatory and Bell dummies	−0.0096**	−0.0069**	−0.0166**
	(−4.92)	(−4.32)	(−6.01)
UNE-P (Hassett and associates, AT&T)			
No dummy variables	−0.0020	−0.0010	−0.0031
	(−1.67)	(−0.74)	(−1.49)
With regulatory dummies	−0.0012	0.0006	−0.0021
	(−1.16)	(−0.45)	(−1.09)
With regulatory and Bell dummies	−0.0009	−0.00004	−0.0012
	(−0.70)	(−0.02)	(−0.50)
CLEC share (FCC, 2002)			
No dummy variables	0.0013	0.0019	0.0006
	(0.57)	(1.07)	(0.17)
With regulatory dummies	0.0008	0.0022	0.0003
	(0.59)	(1.33)	(0.10)
With regulatory and Bell dummies	0.0005	0.0012	−0.0017
	(0.22)	(0.77)	(−0.63)

* Statistically significant at the 10 percent level.

** Statistically significant at the 5 percent confidence level.

a. Figures in parentheses are *t*-statistics.

to 2001–03 rates? Finally, the results using Hassett and associates' measure of the UNE-P rate show no statistically significant relationship between Bell company capital spending and UNE-P rates. Therefore this regression analysis offers little support for the theory that UNE-P rates have affected Bell company capital spending.

A more direct approach to estimating the effect of competition from competitive local exchange carriers (CLECs) on Bell investment would be to insert the actual measure of CLEC market share directly into the investment equation instead of its proxy, the UNE-P rate. If Bell companies adjust their investment plans to variations in the UNE-P rate because they anticipate that low UNE-P rates will lead to more competition, this effect could be modeled more directly by using the actual CLEC share of lines rather than one of the contributors to it, the UNE-P rate. The Federal Communications

Commission publishes data on the entrants' market shares, but in its early reports it withheld the data for a large number of states. In its report on the extent of local competition at the end of 2002, it published these data for thirty-eight states.[4] When I use this measure of competition as a substitute for the various measures of the UNE-P rate, as reported in the last section of table A-3, I find no statistically significant effect of entrant market share on Bell company investment, all things being equal.[5]

Notes

Notes to Chapter One

1. The Telecommunications Act of 1996, 47 U.S.C. 229.

2. The United States had begun to liberalize telecommunications in the 1960s and 1970s, but the 1996 statute was designed to complete the task. A few other countries, such as the United Kingdom and Japan, began to liberalize their telecom sectors in the 1980s, but none other than New Zealand had as extensive a liberalization policy as the United States. New Zealand deregulated telecommunications in 1989 but has since moved back in the direction of regulation.

3. Most notably, Reed Hundt, chairman of the FCC during the first two years of implementation of the 1996 act, claims that the FCC's decisions during this period "have helped foster the rise of not only the New Economy but the Internet Society as well." Reed E. Hundt, *You Say You Want a Revolution: A Story of Information Age Politics* (Yale University Press, 2000), p. x.

4. As this book goes to press, the courts are reining in the regulators. The most recent decision by the U.S. Court of Appeals, *U.S. Telecom Industry Association, et al.* v. *FCC and USA,* No. 00-1012, D.C. Cir. 2004, essentially ends the most ambitious of the regulators' attempts to promote "competition" through aggressive network-sharing requirements. See the discussion in chapters 2 and 4.

5. For the most stunning evidence of this failure, see the FCC's February 20, 2003, decision on the most important regulatory issue that has arisen in implementing the 1996 act, the network-sharing rules. The five members disagreed openly on virtually every detail in this proceeding and hence delayed the publication of their decision by more than six months, creating enormous uncertainty in the industry. Even the chairman dissented from much of the FCC's final decision in this matter, which has now been reversed by the appellate court decision cited in fn. 4.

6. This regulatory history is detailed in Robert W. Crandall and Leonard Waverman, *Talk Is Cheap: The Promise of Regulatory Reform in North American Telecommunications* (Brookings, 1995).

7. Robert W. Crandall and Leonard Waverman, *Who Pays for Universal Service? When Telephone Subsidies Become Transparent* (Brookings, 2000).

8. Ibid.

9. Hong Kong launched its network "unbundling" policy in 1995, one year before the United States made it a national policy.

10. The first example of such a change occurred in New Zealand, where the Commerce Commission recently rejected a similar policy largely because these policies did not appear to be working in the United States or other countries. New Zealand Commerce Commission, *Telecommunications Act 2001, Section 64 Review and Schedule 3 Investigation into Unbundling the Local Loop Network and the Fixed Public Data Network,* Final Report, December 2003 (www.comcom.govt.nz/telecommunications/llu/finalreport.pdf).

11. Values obtained from www.finance.yahoo.com/.

12. Some claim that as much as $2.5 trillion in equity values were lost. See, for example, "North America: The Telecom Scrap Yard," *DooYooo*, July 25, 2002 (www.dooyoo.co.uk/services/telecommunications_services/telecommunication_service_providers_in_general/_review/383840/ [September 2003]).

13. For an informative discussion of the revenues raised from these auctions, see www.st-andrews.ac.uk/economics/undergrad/honours/_EC4515o9a.pdf.

14. See the discussion of foreign carriers in chapter 9.

15. WorldCom and Global Crossing filed for bankruptcy. Global Crossing's assets have been sold to other companies, but WorldCom is attempting to reorganize as MCI. Qwest did not file for bankruptcy protection, but it dismissed its senior officers. It was forced to delay or restate its 2000, 2001, and 2002 financial results owing to accounting irregularities.

16. For a thorough analysis of these adjustment problems, see Clifford Winston, "U.S. Industry Adjustment to Economic Deregulation," *Journal of Economic Perspectives* 12 (Summer 1998): 89–110.

Notes to Chapter Two

1. For a discussion of the early period of competition in the U.S. telephone industry, see Milton Mueller, *Universal Service: Competition, Interconnection, and Monopoly in the Making of the American Telephone System* (Washington: American Enterprise Institute, 1997); and Gerald Brock, *The Telecommunications Industry: The Dynamics of Market Structure* (Harvard University Press, 1981).

2. See Robert W. Crandall and Leonard Waverman, *Who Pays for Universal Service? When Subsidies Become Transparent* (Brookings, 2000).

3. *MCI Telecommunications Corp.* v. *FCC,* 561 F.2d 365 (D.C. Cir. 1977), *cert. denied,* 434 U.S. 1040 (1978); 580 F.2d 590 (D.C. Cir. 1978), *cert. denied,* 439 U.S. 980 (1978).

4. *North Carolina Utilities Commission* v. *FCC,* 552 F.2d 1036 (1977), *cert. denied,* 434 U.S. 874 (1977).

5. *United States* v. *American Telephone and Telegraph Co.*, Civil Action 74-1698 (D.D.C. November 20, 1974).

6. Modification of Final Judgment, *United States* v. *American Telephone and Telegraph Co.*, 552 F. Supp. 131 (D.D.C. 1982), *aff'd. sub. nom., Maryland v. United States,* 460 U.S. 1001, 103 S. Ct. 1240, 75 L. Ed. 2d 472 (1983). For a discussion of this case, see Steve Coll, *The Deal of the Century: The Breakup of AT&T* (Atheneum, 1986). For a discussion of the AT&T case in the context of Sherman Act enforcement, see Robert W. Crandall, "The Failure of Structural Remedies in Sherman Act Monopolization Cases," *Oregon Law Review* 80, no. 1 (Spring 2001): 109–98.

7. The information-services restriction was lifted after the U.S. Court of Appeals reversed the trial court on this provision of the decree. *U.S.* v. *Western Electric, et al., and Pacific Telesis Group et al.*, 900 F.2d 283 (D.C. Cir.), *cert. denied,* 111 S.Ct. 283 (1990).

8. For an estimate of the cost of this process, see Paul H. Rubin and Hashem Dezhbakhsh, "Costs of Delay and Rent-Seeking under the Modification of Final Judgment," *Managerial and Decision Economics* 16 (1995): 385–99.

9. Although Japan has required NTT to establish two separate local companies and a long-distance company within a single holding company, it has not broken NTT into truly independent companies.

10. See Richard W. Oliver and David T. Scheffman, "The Regulation of Vertical Relationships in the U.S. Telecommunications Industry," *Managerial and Decision Economics* 16 (1995): 327–48.

11. The Telecommunications Act of 1996 (47 U.S.C. 151, et. seq.), sec. 253 (a), passed as S. 652, 104 Cong. 2 sess.

12. The FCC's 2003 regulatory decision on local competition, called the "Triennial Review Order," was announced on February 20, 2003, but the Final Order was not published until August 21, 2003. See FCC, *Report and Order on Remand and Further Notice of Proposed Rulemaking,* Review of the Section 251 Unbundling Obligations of Incumbent Local Exchange Carriers; Implementation of the Local Competition Provisions of the Telecommunications Act of 1996; Deployment of Wireline Services Offering Advanced Telecommunications Capability; Appropriate Framework for Broadband Access to the Internet over Wireline Facilities, CC-Dockets 01-338, 96-98, and 98-147 (August 21, 2003). The rules on "network unbundling" in this order were subsequently overturned by the U.S. Court of Appeals in *U.S. Telecom Industry Association, et al.* v. *FCC and USA,* No. 00-0012 (D.C. Cir. 2004). The FCC was about to publish new rules as this book went to press.

13. The United Kingdom has the longest track record: it opened its local market to competition from cable television companies and a second telephone carrier in 1985. However, this market "opening" did not involve unbundling or the imposition of other wholesale service requirements on incumbents. See Robert W. Crandall and Leonard Waverman, "Entry Strategies in Landline Markets," unpublished working paper, 2004.

14. Telecommunications Act of 1996, sec. 251[c][3].

15. Ibid., sec. 251 [d][2].

16. Implementation of the Local Competition Provisions in the Telecommunications Act of 1996, *First Report and Order*, CC Docket 96-98, 11 FCC Rcd 15499 (1996).

17. *AT&T Corp.* v. *Iowa Utilities Board,* 525 U.S. 366 (1999).

18. Implementation of the Local Competition Provisions of the Telecommunications Act of 1996, *Third Report and Order and Fourth Further Notice of Proposed Rulemaking,* 15 FCC Rcd 3696 (1999).

19. *United States Telecom Association et al.* v. *FCC*, 290 F.3d 415 (D.C. Cir. 2002).

20. See fn. 12.

21. In addition, incumbents must share any fiber-optic path to the final subscriber with entrants if the construction of such a fiber loop replaces the old copper loop, which must still be unbundled.

22. For a discussion of the economics of telecommunications interconnection charges, see Jean-Jacques Laffont and Jean Tirole, *Competition in Telecommunications* (MIT Press, 2000), chaps. 4 and 5. See also Mark Armstrong, "The Theory of Access Pricing and Interconnection," in *Handbook of Telecommunications Economics,* vol. 1, edited by Martin E. Cave, Sumit K. Majumdar, and Ingo Vogelsang (Elsevier/North-Holland, 2002), pp. 295–384.

23. The incumbent companies realized the importance of keeping local termination rates low only after new entrants began positioning themselves in front of Internet service providers to "terminate" incoming minutes of Internet connections. Predictably, this highly successful form of arbitrage led to calls for regulators to limit terminating access charges to voice calls. Eventually, regulators closed the arbitrage opportunity, but the incumbents were given a strong message that low, cost-based termination charges were the best choice, even for them.

24. In chapter 4, I show that the use of simple or "total service" resale has lagged severely because of the availability of the unbundled network platform. For avoided costs, see Joan Marsh, AT&T, "UNE-P vs. 271 LD Entry: What's the Real Trade-Off for the RBOCs?" Ex Parte Submission to the FCC, WC Docket 01-338 (September 25, 2002).

25. The chairman of the FCC at that time, Reed Hundt, has written a book on his tenure at the FCC, *You Say You Want a Revolution* (Yale University Press, 2000), that admits to such a goal in implementing the act.

26. *MCI Telecommunications Corp.* v. *FCC*, 561 F. 2d 365 (D.C. Cir. 1977).

27. U.S. Department of Justice and the Federal Trade Commission, *Horizontal Merger Guidelines* (April 2, 1992; rev. April 8, 1997), sec. 1.51.

28. FCC, *Trends in Telephone Service* (May 2004), table 1.2.

29. In its last report to the FCC as a regulated company, AT&T reported total marketing expenses of $4.16 billion in 1994, a year in which it had $36.88 billion in total revenues. Its marketing expenses were roughly two-thirds of the total operating costs and depreciation of its entire national network. See FCC, *Statistics of Communications Common Carriers, 1994/1995* (1995), pp. 42–43.

30. The 1982 AT&T antitrust decree that spun off the Bell operating companies limited each of the divested Bell companies to in-region long-distance service within 163 separate local access and transport areas. They could not offer service between two local access transport areas (LATAs), such as Los Angeles and San Francisco.

31. See 1996 Telecommunications Act, sec. 271.

32. See CTIA, *Semi-Annual Wireless Industry Survey* (www.ctia.org/news_media/news/index.cfm/AID/10030 [June 2004]).

33. In late 2004 Sprint announced that it would merge with Nextel. If this merger is consummated, the number of national wireless carriers will decline to four.

34. For a detailed review of the issues involved in setting wireless termination charges, see Robert W. Crandall and J. Gregory Sidak, "Should Regulators Set Rates to Terminate Calls on Mobile Networks?" *Yale Journal on Regulation* 21 (2004): 261–314.

35. Ibid.

36. Cable Consumer Protection and Competition Act of 1992.

37. The major regulatory issue facing cable is how the new broadband services will be treated under the Communications Act. In 2002 the FCC concluded that "cable modem service, as it is currently offered, is properly classified as an interstate information service, not as a cable service, and that there is no separate offering of telecommunications service. In addition, [it initiated] a rulemaking proceeding to determine the scope of the Commission's jurisdiction to regulate cable modem service and whether (and, if so, how) cable modem service should be regulated under the law." FCC, *Declaratory Ruling and Notice of Proposed Rulemaking,* In the Matter of Inquiry Concerning High-Speed Access to the Internet over Cable and Other Facilities, Internet over Cable Declaratory Ruling, Appropriate Regulatory Treatment for Broadband Access to the Internet over Cable Facilities, GN Docket 00-185 and CS Docket 02-52 (March 15, 2002). This decision has been reversed on appeal, but it is still in the appellate court. See *Brand X Internet Services* v. *FCC and USA,* No. 02-70518, et al. (9th Cir. 2003).

38. Obviously preparing to pursue a policy of limited regulation for cable modem services, the FCC attempted to assert jurisdiction in March 2002 by defining such services as "information" services, thereby depriving state and local offices of the right to regulate them. See FCC, "FCC Classifies Cable Modem Service as 'Information Service,'" GN Docket 00-185 (March 15, 2002). In October 2003, however, this decision was reversed by the Ninth Circuit's decision in *Brand X Internet Services.*

Notes to Chapter Three

1. Vice President Al Gore was credited with this description of modern communications technology. See, for example, www-tech.mit.edu/V113/N65/gore.65w.html.

2. George Gilder, *Telecosm: The World after Bandwidth Abundance* (New York: Simon & Schuster, 2000).

3. These data are from the Bureau of Economic Analysis (BEA), U.S. Department of Commerce (www.bea.doc.gov/bea/dn/faweb/AllFATables.asp#S3 [August 25, 2002]). They are investment expenditures in current dollars for the "telephone and telegraph" industry. The 2002 and 2003 data are estimated from data published by the FCC, Association for Local Telecommunication Services, and the Cellular Telecommunications Industry Association (CTIA) and from the financial reports of the large long-distance carriers.

4. The indexes in figure 3-2 are calculated from monthly closing prices of individual equities, weighted by initial-period market capitalization. The RBOC index is a weighted average of the common equities of SBC, Bell South, and Verizon. The CLEC index is a weighted average of the equities of Allegiance, Covad, McLeod, Time Warner Telecom, and XO Communications, with weights based on base-period market capitalization. The wireless index is a weighted average of the equities of Leap, Nextel, RCCC, and Sprint PCS, with weights based on base-period market capitalization. The long-distance index is a weighted average of the equities of Sprint and WorldCom. As firms disappear, the weights are adjusted to reflect the intial-period market caps of the survivors.

5. The "long-distance companies" in figure 3-2 are Sprint and MCI-WorldCom. AT&T is excluded because for much of 1996–2004 it was a cable television company and owned one of the largest wireless carriers. Its decline was nearly as great as that shown for the other

two carriers in figure 3-2, as its market capitalization declined from nearly $200 billion in early 2000 to $11.5 billion at the end of June 2004.

6. Yochi J. Dreazen, "Wildly Optimistic Data Drove Telecoms to Build Fiber Glut," *Wall Street Journal Online,* September 26, 2002.

7. It is unclear how much of this reported capital spending was devoted to productive capacity. Much of it may have been spent on office facilities, collocation cages, marketing-related equipment, and the like. For a discussion of this issue, see Larry F. Darby, Jeffrey A. Eisenach, and Joseph S. Kraemer, *The CLEC Experiment: Anatomy of a Meltdown* (Washington: Progress and Freedom Foundation, September 2002), pp. 10–15.

8. CTIA, *Semiannual Wireless Survey* (www.wow-com.com/industry/stats/surveys/ [2003]). The CTIA publishes total investment spending by wireless carriers; all other figures are calculated from reports by publicly traded companies to the Securities and Exchange Commission.

9. Note the rise in the wireless equity index in early 2004 in figure 3-2.

10. These data are from BEA (www.bea.doc.gov/bea/dn/faweb/AllFATables.asp#S3 and www.bea.doc.gov/bea/dn2/gpo.htm [accessed December 2002]). The data are no longer available on the BEA website because of the conversion from SIC to NAICS.

11. BEA has not published data on telecommunications sector output for 2002 and 2003.

12. Between 1995 and 2000, the Federal Reserve Board's industrial production index for durable goods manufacturing rose at an average annual rate of more than 8 percent (www.federalreserve.gov/Releases/G17/ipdisk/ip.nsa [accessed December 26, 2004]).

13. FCC, *Telecommunications Industry Revenues, 2002* (March 2004).

14. The GDP chain-type price index rose by slightly less than 2 percent a year, and the consumer price index by slightly more than 2 percent a year over this period.

15. In the late 1990s, there was a widespread belief that Internet traffic would double every 100 days or even faster. This prediction is generally traced to UUNet or WorldCom, the company that bought UUNet. Reed Hundt, in *You Say You Want a Revolution* (Yale University Press, 2000), p. 224, stated that "data traffic" was doubling every 90 days in 2000. It now appears, however, that Internet traffic has been doubling *annually*, not every 90–100 days. See Andrew M. Odlyzko, "Internet Traffic Growth: Sources and Implications," in *Optical Transmission Systems and Equipment for WDM Networking II*, edited by B. B. Dingel, W. Weiershausen, A. K. Dutta, and K.-I. Sato, Proc. SPIE (International Society for Optical Engineering), vol. 5247 (2003), pp. 1–15 (www.dtc.umn.edu/~odlyzko/doc/networks.html).

16. For details, see www.bls.gov/lpc/home.htm.

17. The "UNE-Platform" is the combination of all the "unbundled" network elements required to provide traditional telephone service. By leasing this platform, entrants do not have to invest in their own facilities. See chapters 4 and 5 for further details.

18. U.S. Bureau of the Census, Current Population Survey, as reported in FCC, *Reference Book of Rates, Price Indices and Household Expenditures for Telephone Service* (July 2004), table 2-1.

19. Ibid.

20. The data in table 3-2 exclude household spending on Internet services provided by Internet service providers (ISPs) and cable television companies.

21. For a review of the literature, see Robert W. Crandall and Leonard Waverman, *Who Pays for Universal Service? When Telephone Subsidies Become Transparent* (Brookings, 2000),

chap. 5; and Lester D. Taylor, *Telecommunications Demand in Theory and Practice* (Hingham, Mass.: Kluwer, 1994).

22. See Taylor, *Telecommunications Demand.*

23. Data on interstate and international service appear in FCC, *Telecommunications Industry Revenues, 2002,* table 9. Between 1995 and 2002, the nominal price per minute fell from 16 cents to 9 cents. For intrastate details, see Bureau of Labor Statistics, *Consumer Price Index, All Urban Households,* "Land-line intrastate toll calls," accessible at www.bls.gov/cpi/home.htm (February 2005).

24. The FCC's annual *Telecommunications Industry Revenues* reports show that the share of Bell company revenues accounted for by interstate and international revenues rose from about 16 percent in 2000 to 20 percent in 2002. There was also a large shift to wireless calling, but much of this was at off-peak prices, which were often zero.

25. The 1996 act's major impact on wireless fell on interconnection rates. The act requires local carriers to pay "reciprocal compensation" for terminating each other's calls. These rates are established through the arbitration-regulation process that guides all local interconnection, and they are generally set at about 0.5 to 0.7 cents a minute. Deemed "local" carriers for this purpose, wireless carriers are therefore subject to similar, low interconnection charges.

26. California is currently trying to reestablish regulation in the form of "consumer protection" rules.

27. In late 2004 Sprint announced its intention to merge with Nextel, reducing the number of carriers to four if the merger is consummated.

28. AT&T divested itself of its wireless operations in 2001, and AT&T Wireless recently agreed to be acquired by Cingular. Sprint still owns its Sprint PCS wireless operations.

29. The CTIA's *Semiannual Wireless Industry Survey* reported 169.5 million wireless subscribers in June 2004. According to the FCC's *Local Telephone Competition: Status as of December 31, 2003,* there were 181.4 million wire-based switched access lines in the country.

30. FCC, *Report and Order on Remand and Further Notice of Proposed Rulemaking,* Review of the Section 251 Unbundling Obligations of Incumbent Local Exchange Carriers; Implementation of the Local Competition Provisions of the Telecommunications Act of 1996; Deployment of Wireline Services Offering Advanced Telecommunications Capability; Appropriate Framework for Broadband Access to the Internet over Wireline Facilities, CC-Dockets 01-338, 96-98, and 98-147 (August 21, 2003). In December 2004, the FCC announced a new set of rules, which have not yet been published in complete detail. See FCC, "FCC Adopts New Rules for Network Unbundling Obligations of Local Phone Carriers" (www.fcc.gov. [December 15, 2004]).

31. FCC, *Notice of Inquiry,* In the Matter of Inquiry Concerning High-Speed Access to the Internet over Cable and Other Facilities, GN Docket 00-185 (September 28, 2000); *Declaratory Ruling and Notice of Proposed Rulemaking,* GN Docket 00-185 (March 15, 2002).

32. FCC, *Notice of Proposed Rulemaking,* In the Matter of IP-Enabled Services, WC Docket 04-36 (March 10, 2004).

Notes to Chapter Four

1. For a detailed analysis of some of the costs of this regulatory policy, see Robert W. Crandall and Leonard Waverman, *Who Pays for Universal Service? When Telephone Subsidies Become Transparent* (Brookings, 2000).

2. See FCC, *Trends in Telephone Service* (May 2004), tables 13.1 and 13.2.

3. For a discussion of this early history, see Gerald W. Brock, *The Telecommunications Industry: The Dynamics of Market Structure* (Harvard University Press, 1981).

4. This was also true in the United Kingdom and Canada. See Crandall and Waverman, *Who Pays for Universal Service?* chap. 3.

5. The market capitalization plus long-term debt and other obligations for the Bell companies and GTE at the end of 1996 was $330 billion, of which I estimate that $41 billion reflected the value of their wireless assets. I use the value per subscriber of AirTouch as the basis for estimating the value of RBOC plus GTE wireless subscribers. These companies had 149.1 million switched subscriber access lines at the end of 1996. These data are from company reports to the Securities and Exchange Commission (SEC) and from www.finance.yahoo.com. At the end of 1996 the "reporting" local exchange companies had $296 billion in plant and 155 million switched access lines, or slightly less than $2,000 in book value of plant per line. See FCC, *Statistics of Communications Common Carriers, 1996/97* (December 1997), table 2.7 and 2.10.

6. For example, Qwest announced the sale of its Directory (Yellow Pages) business for $7.05 billion in August 2002. See "Selling Old-Fashioned Telephone Books Helps Telecoms," *Dow Jones Business News*, August 25, 2002.

7. The incumbents may have been inefficient, thereby providing entrants with some potential margin between rates on existing services and the efficient cost of providing them.

8. FCC, *Telecommunications Industry Revenues, 2002* (March 2004), table 3; FCC, *Trends in Telephone Service*, table 7.1.

9. FCC, *Statistics of Communications Common Carriers*, annual editions.

10. These requirements are spelled out in the 1966 act, sec. 251.

11. Joan Marsh, AT&T, "UNE-P vs. 271 LD Entry: What's the Real Trade-Off for the RBOCs?" Ex Parte Submission to the FCC, WC Docket 01-338 (September 25, 2002).

12. The states were Alabama, Arizona, Colorado, Idaho, Iowa, Mississippi, Montana, New Mexico, North Dakota, South Dakota, and Wyoming. By July 2003, only five states—Idaho, Montana, New Mexico, South Dakota, and Wyoming—had rates of $15 a month or more. See Billy Jack Gregg, "A Survey of Unbundled Network Element Prices in the United States" (National Regulatory Research Institute [NRRI]), periodic issues.

13. For a critique of this policy, see Jerry A. Hausman, "The Effect of Sunk Costs in Telecommunications Regulation," in *Real Options: The New Investment Theory and Its Implications for Telecommunications Economics*, edited by James Alleman and Eli Noam (Hingham, Mass.: Kluwer Academic, 1999); Robert W. Crandall and Jerry A. Hausman, "Competition in U.S. Telecommunications Services: Effects of the 1996 Legislation," in *Deregulation of Network Industries: What's Next?* edited by Sam Peltzman and Clifford Winston (AEI-Brookings Joint Center for Regulatory Studies, 2000).

14. The essential facilities argument derives from antitrust law. The landmark case is *Terminal R.R. Association of St. Louis* v. *United States*, 266 U.S. 17 (1924). But even under antitrust law, these essential facilities are never made available to rivals at cost-based regulated rates. In 2004 the Supreme Court closed the door on use of the Sherman Antitrust Act to seek treble damages from a failure to comply with the 1996 Telecommunications Act. *Verizon Communications Inc.* v. *Law Offices of Curtis* v. *Trinko, LLP,* 124 S.Ct. 872 (2004).

15. As of December 2003, 15.2 million incumbent loops were leased "with switching" (that is, as part of the UNE platform), whereas only 4.3 million loops were leased without

switching. See FCC, *Local Telephone Competition: Status as of December 31, 2003* (June 2004), table 4. This reliance on the UNE platform will now begin to dwindle rapidly after the U.S. Court of Appeals' remand of the FCC's 2003 rules and the commission's decision not to appeal this reversal to the Supreme Court.

16. The UNE-P began when the Bell companies were pressured into guaranteeing access to the entire UNE platform at very low rates in negotiations over merger approvals and Section 271 approvals for entry into long distance. Most notably, Verizon (then Bell Atlantic) was forced to allow access to the entire platform in New York as an FCC condition for approval of its acquisition of Nynex in 1999. As a result, CLECs have amassed 30 percent of all switched access lines in New York, nearly double their share of lines in the rest of the country. FCC, *Local Telephone Competition,* table 7.

17. Ibid., tables 3 and 4.

18. Ibid., tables 3 and 5. I am indebted to Tim Tardiff for this insight.

19. Robert W. Crandall, Allan T. Ingraham, and Hal J. Singer, *Do Unbundling Policies Discourage CLEC Facilities-Based Investment?* Berkeley Electronic Papers, Topics in Economics and Policy Research (University of California at Berkeley, 2004).

20. The NRRI study contains the relevant rates that CLECs must pay to operate UNE loops in the fifty states, plus the District of Columbia. Some of the rates are final, and others were under ongoing negotiation as of the July 2001 NRRI study, but all rates are the best available representation of the UNE cost in the respective state. See Billy Jack Gregg, *A Survey of Unbundled Network Element Prices in the United States* (NRRI, Spring 2001). Excluded from the analysis are states that do not publish their average UNE rates: Alaska, Arkansas, California, Colorado, Florida, Hawaii, Louisiana, Missouri, North Dakota, New Mexico, Ohio, Rhode Island, South Carolina, and South Dakota.

21. For each state, the number of E911 lines and the number of UNE loops for 2000 and 2001 were provided by the regional Bell operating companies (RBOCs) that provide local telephone service in that state. For states where more than one RBOC operates, we subtract the *sum* of the UNE loops across each RBOC in that state from the number of CLEC lines in the E911 database.

22. The E911 data are from twenty-three states and cover 2000 and 2001; the FCC data are from thirty-five states and cover the two years between 2001 and 2002.

23. For a contrary view from a CLEC's perspective, see Z-Tel, "Does Unbundling Really Discourage Facilities-Based Entry? An Econometric Examination of the Unbundled Switching Restriction" (February 2002).

24. James Zolnierek, James Eisner, and Ellen Burton, "An Empirical Examination of Entry Patterns in Local Telephone Markets," *Journal of Regulatory Economics* 19 (2001): 143–59; James Eisner and Dale E. Lehman, "Regulatory Behavior and Competitive Entry," paper presented at the 14th Annual Western Conference, Center for Research in Regulated Industries, San Diego, Calif., June 28, 2001; R. Dean Foreman, "For Whom the Bell Alternatives Toll: Demographics of Residential Facilities-Based Telecommunications Competition in the United States," *Telecommunications Policy* 26 (2002): 573–87.

25. Eisner and Lehman, "Regulatory Behavior and Competitive Entry"; Jaison R. Abel, "Entry into Regulated Monopoly Markets: The Development of a Competitive Fringe in the Local Telephone Industry," *Journal of Law and Economics* 45 (October 2002): 289–316.

26. Martin F. McDermott III, *CLEC: An Insider's Look at the Rise and Fall of Local Exchange Competition* (Rockport, Maine: Penobscot Press, 2002).

27. Shane Greenstein and Michael Mazzeo analyze the effect of such strategies in "Differentiation Strategy and Market Deregulation: Local Telecommunication Entry in the Late 1990s," mimeo, April 2003.

28. Dale E. Lehman and Dennis Wiseman, *The Telecommunications Act of 1996: The "Costs" of Managed Competition* (Hingham, Mass.: Kluwer Academic, 2000).

29. Abel, "Entry into Regulated Monopoly Markets." See also Donald L. Alexander and Robert M. Feinberg, "Entry in Local Telecommunications Markets," *Review of Industrial Organization*, forthcoming. Alexander and Feinberg find that the form of state regulation has no significant effect.

30. Eisner and Lehman, "Regulatory Behavior and Competitive Entry."

31. Ibid.

32. The major competitive access providers (CAPs) are owned by larger carriers (such as AT&T and WorldCom) who do not report their results separately; hence I have no data on their capital spending.

33. These capital spending estimates are derived from the annual reports of forty-three public CLECs. They exclude spending by local operations of AT&T and WorldCom. Other estimates suggest a peak in CLEC investment spending of up to $21 billion in 2000. See, for example, New Paradigm Resources Group, *Measuring the Economic Impact of the Telecommunications Act of 1996* (Chicago, October 2002). See also Association for Local Communications Services, *Annual Reports*, for even higher estimates.

34. Competitive access providers such as Teleport Group, MFS, and Brooks Fiber began to provide long-distance companies with special access services before 1996.

35. This is the market value of the outstanding shares plus the book value of the debt of forty-four publicly traded CLECs, excluding Level 3, which had a large investment in fiber-optics facilities throughout the country. Including Level 3, the total would be $150 billion. Based on www.finance.yahoo.com and author's calculations.

36. This calculation is based on twelve of the twenty-one surviving CLECs in my database. The estimate of value per line is undoubtedly biased upward because it is based on the *book* value of debt plus the market value of equity.

37. See www.finance.yahoo.com and company reports to the SEC.

38. Figure 4-4 shows *end-user* revenues. Total revenues include wholesale revenues, which are intercarrier revenues.

39. This calculation is not very sensitive to assumptions about the cost of capital. Even if the *before-tax* cost of capital were as low as 20 percent, the rate of growth would have had to be 19 percent a year.

40. FCC, *Local Telephone Competition*, table 2.

41. As explained later, a logit regression of CLEC subscription finds no statistically significant effect of total telecom spending by households.

42. TNS telecom data.

43. Using TNS household data for the first and second quarters of 2001, I estimated a regression of the household's local bill on a set of dummy variables for each calling feature—call waiting, voice messaging, number identification, call forwarding, and so forth—and the number of lines. According to the results, the prices of new entrants' calling features were very similar to those offered by the incumbents.

44. A personal history of the CLECs, written by Martin F. McDermott III, a chief marketing officer for two CLECs, admits that the CLECs had little to offer customers other

than 15 to 20 percent lower rates. See McDermott, *CLEC: An Insider's Look*, chaps. 11 and 13.

45. The TNS database, drawn from household bills across the country, identifies the household's local carrier and has local billing information for that carrier, but it does not provide the opportunity set of carriers facing any given household.

46. By December 2001, there was at least one CLEC in zip codes that contained 91.2 percent of the country's households and two or more CLECs in zip codes that contained 82.7 percent of the country's households. FCC, *Local Telephone Competition*, table 15.

47. Ibid., table 2.

48. Robert W. Crandall, "An Assessment of the Competitive Local Exchange Carriers Five Years after the Passage of the Telecommunications Act" (www.criterioneconomics.com/docs/Crandall%20CLEC.pdf [accessed December 26, 2004]).

49. There were thirteen carriers in this group: Allegiance, US LEC, Covad, DSLNet, Elec Communications, Focal, ITC DeltaCom, Mpower, Network Access, North Pittsburgh, PacWest, Primus, and Z-tel. I do not attempt to extend this calculation beyond 2001 revenues because of the widespread failures and the lack of data on switched access lines for the dwindling number of survivors.

50. These carriers were Adelphia Business Systems, CoreComm, CTC, CapRock, and McLeod.

51. These carriers were Allied Riser, Cogent, Cypress, Electric Lightwave, GCI, RCN, XO, Time Warner Telecom, USOL, and Winstar.

52. FCC, *Local Telephone Competition*, table 5.

53. Estimates based on various press releases and the trade press. See also New Paradigm Resources Group, *Measuring the Economic Impact of the Telecommunications Act of 1996*.

54. This is not to say that such entry is impossible, but that the new CLECs have not discovered a successful entry strategy. However, the cable television companies and wireless (cellular) carriers remain important sources of competition.

55. Company financial reports and PACE, *UNE-P Fact Report: January 2003* (www.Pacecoalition.org).

56. For details, see chapter 5 or the FCC's website, www.FCC.gov.

57. In August 2003, Sprint announced that it would also begin offering local service by relying on the UNE-P.

58. The FCC data reported by PACE show that 3.1 million of the 5.7 million UNE-Ps were leased in these two states.

59. Keith S. Brown and Paul R. Zimmerman, "The Effect of Section 271 on Competitive Entry into Local Telecommunications Markets: An Initial Evaluation," unpublished ms. (Washington, 2002).

60. See Nicholas Economides, Katja Seim, and V. Brian Viard, "Quantifying the Benefits of Entry into Local Phone Service," paper presented at a conference at the London Business School, May 2004. They estimate that New York subscribers switching from Verizon to AT&T or MCI saved 4.4 percent and 0.7 percent, respectively, in 1999–2003.

61. See Gregory L. Rosston and Bradley S. Wimmer, "Local Telephone Rate Structures: Before and after the Act," Stanford Institute for Economic Research, Discussion Paper 01-30 (August 2002). Rosston and Wimmer find that the 1996 act has not moved local rates closer to their relative costs, and that local urban rates in 2000 were not driven by the cost of service, but by the desire of regulators to use urban rates to defray the cost of rural service.

62. This was part of the "CALLS" compromise that reduced interstate access charges to their current level of about 0.55 cents a minute for the larger ILECs. For a discussion of this plan, see James Eisner, Jim Lande, and Jim Zolnierek, *CALLS Analysis* (Federal Communications Commission, Industry Analysis Division, Common Carrier Bureau, May 25, 2000).

63. FCC, *Trends in Telephone Service,* table 13.1.

64. Data from the Bureau of Labor Statistics (BLS) (www.bls.gov).

65. FCC, *Local Telephone Competition,* table 2.

66. FCC, *Telephone Industry Revenues, Quarterly Roll-ups* (October 2004).

67. FCC, *Reference Book of Rates, Price Indices, and Household Expenditures on Telephone Service* (July 2004), table 2.6. In this analysis, I assume that the CLECs charged 15 percent less than the incumbents charged for local residential service, but I use the $441 average residential rate anyway because I am applying it to the CLECs' residential and small business subscribers.

68. Since there is no evidence that ILEC rates fell, I do not include any effect of this rate competition on ILEC rates.

69. FCC, *Statistics of Communications Common Carriers, 2003–04* (October 2004), table 2.8.

70. "Declaration of William E. Taylor Regarding Special Access Pricing on Behalf of Verizon," submitted to the FCC, WC Docket 04-313 and CC Docket 01-338 (October 4, 2004).

71. Under these assumptions, it appears that the competitors have replaced about one-third of the special access and private lines that the incumbents would have had in 2003. If the competitors are more efficient than the incumbents in providing these lines—say, 30 percent more efficient—then one-third of the $8.4 billion may be a net savings to the economy.

72. Because most of the entry was predicated on the use of incumbents' facilities, the diversion of customers from incumbents to entrants has not reduced the incumbents' need for capital expenditures to maintain their facilities.

73. Author's tabulation from reports of forty-five carriers to the SEC. This does not include local investments by MCI or AT&T.

74. Association for Local Telecommunication Services (ALTS), *The State of Local Competition, 2004* (July 2004). The ALTS data include expenditures by Level 3, AT&T, MCI, and some cable companies.

75. Even if the assets had infinite life, a 15 percent before-tax capital cost on $55 billion would require $8.25 billion in annual capital charges.

76. Morgan Stanley Dean Witter, *Telecommunications Services,* Industry Report (July 19, 2000); Mark Kastan and Daniel Reingold, *Telecom Services—Local* (Merrill Lynch, June 3, 1999).

77. Morgan Stanley, *Telecommunications Services.*

78. Ironically, the availability of the UNE platform at artificially low rates has made it more difficult for entrants with their own facilities to market their services. The proliferation of marketing programs by UNE-P carriers increases the marketing costs per new customer for other carriers who may actually offer a new service or innovation. Remarks of John Malone, Eastern Management Group, at the CATO Institute, September 24, 2003.

79. Author's tabulation based on annual reports to the SEC by thirty-two CLECs and employment data from BLS.

80. I assume that resale adds nothing to telecom output, that revenues from services using UNE loops contribute to output an equivalent of 30 percent of the revenue share from such services, and that revenues of facilities-based CLECs contribute to output in full proportion to their share of industry revenues, all based on FCC revenue data. I adjust labor input by reducing it by the share of CLEC employment in total fixed-wire industry employment in 2002. I do not adjust ILEC employment.

81. This assessment does not take into the account the potentially large economic costs of denying the Bell companies the ability to compete in interLATA long distance between 1996 and 2002. Long-distance rates would surely have been much lower over this period if the Bell companies had been allowed to enter. See chapter 6 for an estimate of this additional cost to consumers.

82. I discuss broadband services in chapter 8. I do not, however, attempt to measure the extent of innovation in services to small or medium businesses or any potential increase in service quality to these customers that has resulted from entry.

Notes to Chapter Five

1. I address wireless-wireline substitution in chapter 7.

2. Verizon acquired GTE in 2000. SBC acquired Ameritech, Pacific Telesis, and Southern New England Telephone in 1997–99. Qwest acquired U S West in 2000.

3. *AT&T Corp. et al.* v. *Iowa Utilities Board et al.*, 525 U.S. 366 (2000).

4. The cable companies have outperformed the Bell companies, but their equity prices have been much more volatile over this period.

5. In 1991 they succeeded in overturning the "information services" ban of the 1982 decree. Without being able to participate in long-distance services and manufacturing activity (including engineering design), however, this victory proved to be of limited value.

6. *U.S. Telecom Industry Association et al.* v. *FCC and USA,* No. 00-1012, D.C. Cir. 2004.

7. Technically, the beta measure is the *covariance* of the company's equity price with the overall market. A beta of more than 1.0, therefore, suggests that investors view the stock as more risky than the average for the entire market. A beta of 0.50 suggests the contrary, namely, that the security is less risky than the average for the entire market.

8. See chapter 9 for a comparison with the incumbent telephone companies in other developed countries.

9. The wholesale loop rates have been lowered in twenty-five of the forty-eight mainland states since 2001. See Billy Jack Gregg, *A Survey of Network Unbundled Element Prices in the United States* (National Regulatory Research Institute), periodic issues.

10. Similar results were obtained by Allan Ingraham and J. Gregory Sidak in "Mandatory Unbundling, UNE-P and the Cost of Equity: Does TELRIC Pricing Increase Risk for Incumbent Local Carriers?" *Yale Journal on Regulation* 20 (2003): 389–406.

11. Growth in the Standard & Poors 500 Index, December 31, 1996, to December 31, 2003 (http://finance.yahoo.com).

12. FCC , *Telecommunications Industry Revenues, 2002* (March 2004), table 3 (adjusted).

13. Anna Maria Kovacs, Kristin L. Burns, and Gregory S. Vitale, *The Status of 271 and UNE-Platform in the Regional Bells' Territories* (Philadelphia: Commerce Capital Markets, November 8, 2002).

14. Data from FCC, *Local Telephone Competition: Status as of December 31, 2003* (June 2004), table 1. The current population survey for November 2001 showed that 1.2 percent of households had a wireless telephone but no fixed wireline phone. Owing to technical problems, the results from CPS surveys in 2002 and 2003 were somewhat different, but the February 2004 survey found that 6 percent of households had only wireless service. See chapter 7.

15. A new carrier, Vonage, has begun to offer VoIP services over cable networks, placing substantial pressure on cable companies to launch or expand their own VoIP services. These developments have in turn led the FCC to begin a regulatory proceeding to determine whether VoIP services will be treated like other voice services, thereby requiring the payment of switched access charges, universal service fees, and excise taxes or 911 services and law-enforcement access for wire-tapping purposes.

16. FCC, *Statistics of Communications Common Carriers, 2003–04* (October 2004), table 2.8.

17. Intrastate access charge revenues declined by less than 20 percent between 1995 and 2002, while switched interstate revenues fell by 60 percent. FCC, *Statistics of Communications Common Carriers, 1995 and 2002–03.* The 2003–04 edition does not break out intrastate from interstate revenues.

18. TNS data reported by the FCC, *Reference Book of Rates, Price Indices and Household Expenditures for Telephone Service* (2004), table 2.6.

19. See www.timewarnercable.com/corporate/products/digitalphone/default.html (July 21, 2004).

20. "Crossed Wires," *Economist,* February 15–21, 2003, p. 60.

21. "Comcast Anounces VoIP Service," *TechWeb News,* May 26, 2004 (www.networking pipeline.com/news/21100406 [July 21, 2004]).

22. FCC, *Telecommunications Industry Revenues, 2002* (March 2004). The 2003 data are based on the "roll-up" of quarterly data reported by the FCC.

23. See, for example, Viktor Shvets and others, *1Q03 Results–A Light at the End of the Tunnel?* (New York: Deutsche Bank, April 29, 2003); John C. Hodulik, *Wireline Telecom Play Book* (New York: UBS Securities, July 8, 2004).

24. Robert D. Willig, William H. Lehr, John P. Bigelow, and Stephen B. Levinson, *Stimulating Investment and the Telecommunications Act of 1996,* paper filed by AT&T in FCC Docket 01-338 (October 11, 2002). See also Consumer Federation of America, "Competition at the Cross Roads: Can Public Utility Commissions Save Local Competition?" undated ms.

25. Gregg, *A Survey of Network Unbundled Element Prices.*

26. Only one state, Nebraska, raised UNE rates.

27. Southern New England Telephone was acquired by SBC. Rochester Telephone became Frontier, which was subsequently acquired by Global Crossing. United Telephone is part of Sprint, the large national long-distance company.

28. Data from Broadwing's Annual Report to the Securities and Exchange Commission, Form 10K.

29. I include Qwest with the other three companies here because virtually all of its market value is in its local telephone systems formerly owned by U S West. Its long-distance assets have little or no value. See chapter 6.

30. Sprint also has a large presence in local telephony. It pursued a rather conservative business strategy, expanding only into wireless but avoiding the temptation to grow by merger even though it was not constrained by the 1996 act. Qwest built its long-distance network aggressively, acquiring U S West along the way even though U S West was subject to the interLATA restrictions in the act.

31. Through wholesale and retail arrangements.

Notes to Chapter Six

1. Much more detail may be found in Robert W. Crandall, *After the Breakup: U.S. Telecommunications in a More Competitive Era* (Brookings, 1991).

2. Ranking the long-distance carriers has become a perilous enterprise with the spread of "bundled" plans by local, long-distance, and wireless carriers. Many of these plans allow unlimited calling or free calling during specified periods. As a result, it is increasingly difficult to separate "long-distance" charges from local charges or even broadband charges.

3. Before the AT&T divestiture, long-distance charges reflected an arbitrary division of the company's costs between the interstate and state jurisdictions. See Robert W. Crandall and Leonard Waverman, *Talk is Cheap: The Promise of Regulatory Reform in North American Telecommunications* (Brookings, 1996), chap. 5. This revenue division was designed to keep local rates low. When AT&T was broken up, the FCC was forced to establish formal access charges, which are paid by AT&T and other long-distance carriers to the local carriers, including the divested Bell operating companies.

4. FCC, *In the Matter of Access Charge Reform, et.al.*, Sixth Report and Order in CC Docket 96-262 and 94-1, Report and Order in CC Docket 99-249, Eleventh Report and Order in CC Docket 96-45 (May 31, 2000).

5. For estimates of the cost of this policy, see Robert W. Crandall and Leonard Waverman, *Who Pays for Universal Service? When Telephone Subsidies Become Transparent* (Brookings, 2000).

6. Jeffrey H. Rohlfs and J. Gregory Sidak, "Exporting Telecommunications Regulation: The U.S.-Japan Negotiations on Interconnection Pricing," Working Paper 02-3 (AEI-Brookings Joint Center for Regulatory Studies, March 2002).

7. Obviously, interconnection costs were likely much higher in the 1970s and 1980s than today, but not nearly as high as the implicit rate built into the settlements process before 1984.

8. "Special Access" revenues for the Bell companies have increased from $3.07 billion in 1996 to $12.84 billion in 2002, while interstate switched access revenues have declined from $9.41 billion to $4.35 billion. FCC, *Statistics of Communications Common Carriers,*1996–97 and 2002–03 editions. This shift occurred mainly in originating traffic. The alleged unlawful diversion of traffic discussed here refers to terminating traffic.

9. Tim McElligott, "AT&T Airing Industry's Dirty Laundry with New MCI Fraud Allegations," *Telephony Online*, July 29, 2003 (http://telephonyonline.com/ar/telecom_att_strings_line/ [August 25, 2003]).

10. Data on interstate revenue per conversation minute are not available before 1992, and terminating access minutes are not available for 2003. I assume that the elasticity of demand with respect to interstate and international rates is –0.75 and that the elasticity with respect to GDP and population is 1.0.

11. If the estimate of diversion of switched interstate access minutes is correct, the FCC's reported interstate and international revenues per minute are overstated for the period after 1995. By 1999 this overstatement would be 12 percent, or about 0.9 cents per interstate minute and 1.3 cents per interstate and international minute.

12. The Herfindahl-Hirschman index of concentration for long-distance carriers fell from 4074 in 1992 to 2315 in 2001. See FCC, *Statistics of the Long-distance Telecommunications Industry*, 4th quarter 1998 and 2003.

13. The AT&T decree divided the country into 161 local access and transport areas (LATAs). The divested Bell companies were not allowed to provide service between LATAs if the service originated in their home region.

14. It is puzzling that AT&T Wireless launched this new pricing given that the nationwide bundles would compete with its own fixed-line long-distance business. One would have thought that Cingular, Verizon, or Nextel would have been the first to offer such plans, not AT&T.

15. See the CTIA website, www.ctia.org/news_media/news/index.cfm/AID/10030. Wireless revenues nearly doubled in the three years following AT&T's introduction of its One-Rate Plan, and roaming revenue growth slowed in 1998–2000, then declined in 2001.

16. Recall that these rates may have actually been somewhat lower because of an understated denominator in the FCC's calculation.

17. For similar conclusions, see the analysis of changes in long-distance rates and effect of competition on rates in Paul W. MacAvoy, *The Failure of Antitrust and Regulation to Establish Competition in Long-Distance Telephone Services* (Cambridge, Mass.: MIT Press, 1996); and William E. Taylor and Lester D. Taylor, "Postdivestiture Long-Distance Competition in the United States," *American Economic Review: Papers and Proceedings* (May 1993), pp. 185–90.

18. Large business customers can use private lines or "special access" to avoid originating switched access charges, but such options are not available for incoming calls except through the use of 800 numbers. For this reason, terminating access minutes are often used as a proxy for *conversation* minutes.

19. FCC, *Eighth Annual CMRS Competition Report* (July 2003), appx. D, table 9.

20. Ibid.

21. According to the CTIA *Semiannual Wireless Survey,* there were 140.8 million subscribers at the end of 2002 and 128.4 million at the end of 2001. I therefore assume that the average subscriber level over the year was 134.6 million. The difference in minutes of use is approximately 300 a month, or 3,600 a year. Thus the total estimated increase is 484.6 billion minutes.

22. These estimates are derived from TNS data and are published in FCC, *Trends in Telephone Service* (May 2004), table 11.4.

23. The Bell company lines (RBOCLINES) variable is the sum of switched access lines in states in which the Bell companies have received Section 271 approval, as reported in the FCC's *Statistics of Communications Common Carriers* for December 31, 2001. Average wireless minutes of use per subscriber per month (WIRELESSMIN/MO) are interpolated from semiannual reports from the CTIA. RBOCLINES was lagged to allow for time for Bell

companies to develop and market long-distance service plans. The best fit was found in an equation in which RBOCLINES is lagged three months. The equation was estimated with and without a time trend. Because the wireless minutes variable (WIRELESSMIN/MO) and the time trend are collinear, the statistical significance of each is reduced substantially when both are included; therefore I use the equation without a time trend. The magnitude and statistical significance of the Bell lines variable are not affected by the inclusion of the time trend, however. The results are:

$$\text{CPI-LDReal} = 108.8 - \underset{(t = -11.26)}{0.073}\ \text{WIRELESSMIN/MO} - \underset{(t = -2.69)}{0.0402}\ \text{RBOCLINES}(-3)$$

where CPI-LDReal is the consumer price index for residential long distance deflated by the overall CPI. R_2 (adj.) = 0.966.

24. Assuming that the price elasticity of demand is –0.7.

25. To the extent that this policy kept inefficient carriers alive, much of this transfer was also a "deadweight" loss to the economy. In addition, lower long-distance rates would have induced greater consumption of long-distance services. The increase in consumer surplus from greater long-distance calling would be about $112 million a year.

26. In July 2003, Liberty Media agreed to purchase the rest of QVC's stock for a price that translates into a market capitalization of $14 billion.

27. As this book goes to press, SBC is in the process of acquiring AT&T, and MCI is being acquired by Verizon.

28. The total capital expenditure estimates in figure 6-4 are based company reports to the SEC for AT&T, WorldCom, Sprint, Qwest, and Global Crossing. I have attempted to eliminate the capital spending by Qwest on U S West facilities, by AT&T on its cable facilities, and by Sprint and AT&T on their wireless networks.

29. Quarterly Report to the SEC, Form 10Q, May 10, 2004.

Notes to Chapter Seven

1. The FCC reported 181.4 million fixed-wire switched access lines in the country at the end of 2003. See FCC, *Local Competition: Status as of December 31, 2003,* table 1. At this time there were 158.7 million wireless subscribers, according to the Cellular Telecommunications Industry Association (CTIA), *Semiannual Wireless Industry Survey* (http://files.ctia.org/pdf/CTIA_Semiannual_Survey_YE2003.pdf [2003]). Since fixed-wire lines are declining and wireless subscribers are still increasing at an annual rate of 10 percent or more, the number of wireless subscribers should exceed the number of fixed-wire switched access lines sometime in 2005.

2. AT&T Wireless was purchased by Cingular in October 2004, and Sprint and Nextel announced merger plans in December 2004.

3. The growth of wireless has sharply reduced payphone revenues. Between 1998—the year in which AT&T Wireless began to offer a nationwide calling plan—and 2002, payphone revenues declined by more than 50 percent. FCC, *Telecommunication Industry Revenues 2002* (March 2004), table 2.

4. FCC, *Local Competition: Status as of December 31, 2003,* table 1.

5. Ibid., table 2.

6. One would have expected these competitive services to remain with AT&T, but apparently AT&T did not pursue them aggressively because it did not believe that wireless services would be an important business.

7. Omnibus Budget Reconciliation Act of 1993, Public Law 103-66, Title VI.

8. For evidence of the counterproductive effect of regulation on wireless rates, see Philip M. Parker and Lars-Hendrik Roller, "Collusive Conduct in Duopolies: Multimarket Conduct and Cross-Ownership in the Mobile Telephone Industry, *Rand Journal of Economics* 28 (Summer 1997): 304–22.

9. Denise Pappalardo and Jim Duffy, "Cingular, AT&T Wireless Face Hurdles," *NetworkWorldFusion,* February 23, 2004 (www.nwfusion.com/news/2004/0223attwireless.html [August 1, 2004]).

10. The FCC has also imposed number portability and E911 requirements on the carriers.

11. Cellular Telephone Industry Association (CTIA), *Semiannual Wireless Industry Survey.*

12. Ibid.

13. The average cellular bill declined from $96.83 a month in the second half of 1987 (the first date for which data are available) to $39.43 in the second half of 1998 and has risen steadily since then to $49.91 a month in the second half of December 2003 despite sharply declining rates. See CTIA, *Semiannual Wireless Industry Survey.*

14. FCC, *Eighth Annual CMRS Competition Report* (July 2003), table 9.

15. For details, see www1.sprintpcs.com/explore/ExploreHome.jsp.

16. In 2003 Verizon Wireless began to roll out a 300- to 500-kilobit per second service in major metropolitan areas.

17. Morgan Stanley, Wireless Data Report, as cited in FCC, *Eighth Annual Competition Report,* 418.

18. For details, see www.fcc.gov/auctions/default.htm?job=auctions_home.

19. For a review of the confusion created by this FCC policy in auctioning the C-band frequencies and the Supreme Court's decision to uphold an appellate court decision that prevented the FCC from requiring a defaulting bidder, Nextwave, to return its spectrum to the Commission, see *FCC* v. *Nextwave Personal Communications, Inc.,* 123 S. Ct. 832 (2003).

20. See www.nwfusion.com/edge/news/2003/0804qwest.html.

21. As noted earlier, Sprint's merger with Nextel may reduce this number to four if it is consummated in 2005.

22. FCC, *Eighth Annual CMRS Competition Report.*

23. Data from FCC, *Telecommunications Industry Revenues 2002* (March 2004), tables 9 and 10. As discussed in chapter 6, this decline may be understated because of the understatement of interstate switched access minutes. Nevertheless, wire-based long-distance rates have not fallen nearly as rapidly as cellular rates. The FCC estimate of interstate revenues per interstate conversation minute is available only through 2002.

24. Bureau of Labor Statistics (BLS), *Consumer Price Index, All Urban Consumers.* The current CPI for landline long-distance service is only available for months beginning in December 1997. A weighted average of the decline in interstate revenues per minute from the FCC's *Telecommunications Industry Revenues* report and the rate of decline in intrastate charges registered in the BLS *Consumer Price Index* with weights of two-thirds and one-third, respectively, declines at an average rate of 8.9 percent a year between 1996 and 2002.

25. BLS, *Consumer Price Index;* and FCC, *Trends in Telephone Service* (May 2004), table 13.3.

26. Mark Rodini, Michael R. Ward, and Glenn A. Woroch, "Going Mobile: Substitutability between Fixed and Mobile Access," *Telecommunications Policy* 27 (June 2003): 457–76. A more recent article obtains similar results using a pooled times-series, cross-section sample of aggregate subscriber data from fifty-six countries: see Gary Madden and Grant Coble-Neal, "Economic Determinants of Global Mobile Telephony Growth," *Information Economics and Policy*, available online December 4, 2003.

27. FCC, *Local Telephone Competition: Status as of December 31, 2003* (June 2004), table 13.

28. The price variable was the price of residential service in the largest city in the state as published by the FCC, *Reference Book of Rates, Price Indices, and Expenditures for Telephone Service* (July 2004).

29. The estimated cross-price elasticity was 0.18 for the local fixed-wire rate.

30. FCC, *Telephone Subscribership in the United States (Data through March 2004)* (August 2004), table 1.

31. Ibid., p.2, fn. 2.

32. Steve Kirkeby, J. D. Power and Associates, paper presented to the 31st Telecommunications Conference, KMB Video, St. Petersburg Beach, Florida, May 7, 2003.

33. Ernst & Young/Primetrica, *Mobile Wireless-Primary Fixed Line Substitution* (2003).

34. The price variable was the price of residential service in the largest city in the state as published by the FCC, *Reference Book.*

35. The HHI will rise if Sprint's meger with Nextel is completed.

36. For an early example of this methodology, see Eric B. Lindenberg and Stephen A. Ross, "Tobin's *q* Ratio and Industrial Organization," *Journal of Business* 54 (January 1981): 1–32.

37. Some of the variance may reflect the differential costs of moving the former users of such spectrum to new frequencies, as required by the FCC. Moreover, the later "reauctions" were bedeviled by legal uncertainties concerning previous defaults by winning bidders.

38. See "Cingular Wireless and NextWave Telecom Agree to Terms for Spectrum Licenses," *PRNewsWire,* August 5, 2003 (http://prnewswire.com/cgi-bin/stories.pl?ACCT=104&STORY=/www/story/08-05-2003/000199...).

39. CTIA, *Semiannual Wireless Industry Survey.*

40. Ibid.

41. Data derived from the carriers' financial statements.

42. The market caps of all three carriers have been driven up recently by Cingular's January 2004 bid for AT&T Wireless.

43. Sprint has offered $35 billion to purchase Nextel, which has about 15.3 million subscribers, or about $2,300 per subscriber.

44. For an analysis of this problem, see Robert W. Crandall and J. Gregory Sidak, "Should Regulators Set Rates to Terminate Calls on Mobile Networks?" *Yale Journal on Regulation* 21 (Summer 2004): 261–314.

45. In a regression analysis of 1999–2001 wireless penetration across countries, using ITU data, I found that penetration is positively and significantly related to the peak three-minute local wireline rate in the country and inversely related to the average price of cellular service.

46. The minutes for European countries reflect only outgoing calls, while the U.S. data include both outgoing and incoming cellular minutes. Depending on the mix of traffic, this

could result in a substantial overstatement of the difference between the United States and Europe, but the large gap shown in table 7-4 could not possibly be due simply to this anomaly. Even if one doubles the minutes for EU countries, surely an overadjustment, European use is still far less than U.S. use for the average cellular subscriber.

Notes to Chapter Eight

1. U.S. Department of Commerce, National Telecommunications and Information Administration, *A Nation Online,* 2002. See also www.zakon.org/robert/internet/timeline/.

2. Census Bureau data from www.census.gov, as reported by the National Telecommunications and Information Administration, U.S. Department of Commerce.

3. Jupiter Research estimate as reported by Alex Goldman, "Top 22 U.S. ISPs by Subscriber: Q1 2004," *ISP-Planet,* May 20, 2004 (www.isp-planet.com/research/rankings/usa.html [August 15, 2004]). At the end of 2002, there were 111.3 million households in the United States, according to the U.S. Census Bureau, *Historical Income Tables: Households* (www.census.gov/hhes/income/histinc/h0601.html [December 26, 2004]).

4. For a discussion of the rate of price decline in such equipment, see Charles L. Jackson, "Wired High-Speed Access," in *Broadband: Should We Regulate High-Speed Internet Access?* edited by Robert W. Crandall and James H. Alleman (Brookings, 2002).

5. The Federal Communications Commission currently defines "high speed" as a service providing at least 200 kilobits per second in one direction. See FCC, *High-Speed Services for Internet Access: Status as of December 31, 2003* (June 2004), p.1, fn. 1. This is a relatively slow speed for modern broadband connectivity. This brief technical description is based on material in Jackson, "Wired High-Speed Access."

6. Recall the discussion of the FCC's 2003 "Triennial Review" decision in chapter 2.

7. See http://hns.getdway.com/ and www.starband.com.

8. In its 2003 10-K Annual Report to the Securities and Exchange Commission (p. 3), Sprint stated that it was ending its "pursuit of a residential fixed wireless strategy" using these licenses. MCI reported in its 2003 10-K (p. 38) that it had sold its MMDS licenses and related assets in June 2003 after posting a 1.4 billion loss on the business.

9. A longer-range version, popularly known as WiMAX, is also under development, but it is not yet in widespread commercial use. See Telephony's "Complete Guide to WiMAX: The Business Case for Service Provider Deployment," *Telephony OnLine.com,* June 2, 2204 (http://telephonyonline.com/ar/telecom_telephonys_complete_guide/index.htm).

10. See www.elektrosmog.nu/; and Rob Flickenger, *Building Wireless Community Networks,* www.oreilly.com/catalog/wirelesscommnet/index.html (2001). There is a growing interest in using "unlicensed" spectrum for a variety of communications, including broadband. For an enthusiastic description of the prospects for shared, unlicensed spectrum, see, for example, "On the Same Wavelength," *Economist,* August 14, 2004. For a critical view, see Thomas Hazlett and Matthew Spitzer, *In the Matter of an Interference Temperature Metric to Quantify and Manage Interference and to Expand Available Unlicensed Operation in Certain Fixed, Mobile, and Satellite Frequency Bands,* Docket ET 03-237 (FCC, April 5, 2004).

11. Ephraim Schwartz, "Verizon Races toward Wireless Broadband," *InfoWorld,* May 7, 2004 (www.infoworld.com/article/04/05/07/19FEextenddev_1.html [August 15, 2004]).

12. FCC, *High-Speed Services for Internet Access: Status as of December 31, 2003* (June 2004).

13. John Haring and Jeffrey Rohlfs, "The Disincentives for Broadband Deployment Afforded by the FCC's Unbundlimg Policies," in Comments of the High Technology Broadband Coalition, submitted in FCC, *Review of the Section 251 Unbundling Obligation of Incumbent Local Exchange Carriers* (April 5, 2003).

14. Korea is far ahead of every other country in broadband penetration because of its policies regarding broadband infrastructure and the large share of households in recently built high-rise apartments. See Jerry Hausman, "Internet-Related Services: Results of Asymmetric Regulation," in Crandall and Alleman, *Broadband.*

15. ECTA data may be accessed at www.ectaportal.com/html/index.php (August 15, 2004).

16. The most recent of these entreaties comes from Reed Hundt in a speech delivered at the New America Foundation, December 10, 2003. Hundt, FCC chairman in the 1990s, advocates a government policy that will extend a "Big Broadband" service of 10–100 megabits per second to all households and businesses in the United States.

17. This experiment is described in Hal Varian, "The Demand for Bandwidth: Evidence from the INDEX Project," in Crandall and Alleman, *Broadband.*

18. Paul Rappoport, Donald J. Kridel, and Lester D. Taylor," The Demand for Broadband: Access, Content and the Value of Time," in Crandall and Alleman, *Broadband.*

19. Paul Rappoport and others, *Residential Demand for Access to the Internet*, Working Paper (University of Arizona, Spring 2001).

20. Declaration of Robert W. Crandall and J. Gregory Sidak, *In the Matter of SBC Petition for Expedited Ruling That It Is Non-Dominant in Its Provision of Advanced Services and for Forbearance from Dominant Carrier Regulation of Those Services*, FCC, 2001.

21. See Jung Hyun Kim, Johannes M. Bauer, and Steven S. Wildman, "Broadband Uptake in OECD Countries," paper presented at the 31st Research Conference on Communications, Information and Internet Policy, Arlington, Va., September 2003. The authors find that the price of broadband has little effect on penetration across OECD countries. This may be because the data across the thirty countries in their sample are not comparable.

22. The FCC resolved one of the major issues in October 2004 when it announced it would forbear from requiring wholesale unbundling of facilities for broadband services under Section 271 of the 1996 act. FCC, *Memorandum Opinion and Order,* WC Docket 01-338, WC Docket 03-235, WC Docket 03-260, WC Docket 04-48 (October 27, 2004).

23. This discussion is adapted from Robert W. Crandall, Robert W. Hahn, and Timothy J. Tardiff, "The Benefits of Broadband and the Effect of Regulation," in Robert W. Crandall and James Alleman, *Broadband.* See also Robert W. Crandall and Charles L. Jackson, "The $500 Billion Opportunity: The Potential Economic Benefit of Widespread Diffusion of Broadband Internet Access," in *Down to the Wire: Studies in the Diffusion and Regulation of Telecommunications Technologies,* edited by Allan Shampine (Haupaugge, N.Y.: Nova Science, 2003).

24. The most important source of this uncertainty is the potential for the FCC to invoke another provision of the 1996 act, section 271, to require network sharing that the courts have not allowed the FCC to mandate under Section 251. Section 271 involves the conditions required to allow the Bell companies to provide in-region interLATA services.

25. FCC, *In the Matters of Deployment of Wireline Services Offering Advanced Telecommunications Capability*, Second Report and Order, CC Docket 98-147 (November 9, 1999), para. 21. Indeed, while ILECs are required to provide services to ISPs at regulated prices, the FCC has not even required that cable television providers offer ISPs access at *any* price.

26. See, for example, the complaint filed at the FCC in 2004: *Earthlink, Inc., Complainant,* v. *SBC Communications, Inc. and SBC Advanced Solutions, Inc.,* May 2004.

27. The Ninth Circuit recently overturned this FCC ruling. See chapter 2.

28. Thomas W. Hazlett, "Regulation and Vertical Integration in Broadband Access Supply," in Crandall and Alleman, *Broadband.*

29. An exception is Christopher S. Yoo, "Would Mandating Broadband Network Neutrality Help or Hurt Competition? A Comment on the End-to-End Debate," *Journal on Telecommunications and High Technology Law* 3. vol. 1 (2004): 23–68.

30. For a discussion of some of these earlier precedents, see Crandall, Hahn, and Tardiff, "The Benefits of Broadband and the Effect of Regulation." See also Robert W. Crandall, "The Remedy for the 'Bottleneck Monopoly' in Telecom: Isolate It, Share It, or Ignore It," *University of Chicago Law Review* 2005 (forthcoming).

31. *In the Matter of America Online, Inc. and Time Warner Inc.*, FTC Docket C-3989, Agreement Containing Consent Orders; Decision and Order, 2000 WL 1843019 (FTC) (proposed December 14, 2000) ("FTC Consent Agreement"). For a discussion of the problems posed by the merger, see Gerald R. Faulhaber "Network Effects and Merger Analysis: Instant Messaging and the AOL-Time Warner Case, *Telecommunications Policy* 26 (June 5–6, 2002): 311–33; and Faulhaber, "Access and Network Effects in the 'New Economy': AOL-Time Warner," in *The Antitrust Revolution,* edited by John E. Kwoka Jr. and Lawrence J. White, 4th ed. (Oxford University Press, 2003).

32. From the beginning of 2001 through June 30, 2004, AOL-Time Warner (now Time Warner) had lost about 50 percent of its value whereas the S&P 500 was down only 14 percent.

33. See Robert W. Crandall and Clifford Winston, "Does Antitrust Policy Improve Consumer Welfare? Assessing the Evidence," *Journal of Economic Perspectives* 17 (Fall 2003): 3–26.

34. In 1994 only 2 percent of households had access to the Internet. See National Science Foundation, Directorate for Social, Behavioral, and Economic Sciences, "Complex Picture of Computer Use in Home Emerges," NSF 00-314 (March 31, 2000).

35. Mark A. Lemley and Lawrence Lessig, "The End of End-to-End: Preserving the Architecture of the Internet in the Broadband Era," *Berkeley School of Law and Economics Working Papers* 2 (Fall 2000): art. 8.

36. See, for example, Kevin Werbach, "A Layered Model for Internet Policy," *Journal on Telecommunications and High Technology Law* 1, no. 1 (2002): 37–68.

37. For an excellent overview of the theory for calculating the impact of delay, see Jerry Hausman, "Valuing the Effect of Regulation on New Services in Telecommunications," *Brookings Papers on Economic Activity: Microeconomics* (1997): 1–38. For an estimate of the potential losses from delaying broadband, see Robert Crandall and Charles L. Jackson, *The $500 Billion Opportunity* (www.criter/oneconomics.com/pubs/articles_crandall.php [2001]).

38. Ibid.

39. Ibid.

40. Niraj Gupta, Jack B. Grubman, and Kara Swenson, "The Battle for the High-Speed Data Subscriber: Cable vs. DSL" (Salomon, Smith, Barney, August 20, 2001).

41. Some evidence on the effects of UNE-P regulation was provided in chapter 5.

42. As already mentioned, the FCC subsequently eliminated the possibility of regulating incumbents' broadband services under Section 271 of the 1996 act. See fn. 22.

43. Figure 8-6 shows that DSL had reached a plateau of slightly less than 80 million "addressable locations." There are about 100 million telephone households in the United States and 20 million small businesses.

44. Its stock market price declined by about two-thirds between September 2003 and June 2004, undoubtedly in response to the court decision.

45. FCC, *High-Speed Services for Internet Access: Status as of December 31, 2003*, table 5.

46. Debra J. Aron and David E. Burnstein, "Broadband Adoption in the United States: An Empirical Analysis," March 2003 (http://papers.ssrn.com/sol3/papers.cfm?abstract_id=386100). For a similar conclusion in an analysis of broadband penetration across 135 countries, see Martha Garcia-Murillo and David Gabel, "International Broadband Deployment: The Impact of Unbundling," paper presented at the 31st Research Conference on Communications, Information and Internet Policy, Arlington, Va., September 2003. Privatization matters, but network unbundling does not.

47. Michelle S. Kosmidis, "Deployment of Broadband Infrastructure in the EU: Is State Intervention Necessary?" paper presented at the Telecommunications Policy Research Conference, Alexandria, Va., September 2002.

48. Data obtained from the ECTA website, http://www.ectaportal.com.

49. See www.broadbandweek.com/news/020715/020715_telecom_sbc.html.

50. See ibid.

51. Christopher Rhoads, "Bringing Fiber Home," *Wall Street Journal*, August 19, 2004, p. B1.

52. FCC, *High-Speed Services for Internet Access: Status as of December 31, 2003*, table 3.

Notes to Chapter Nine

1. Data from www.ofcom.org.uk/static/archive/oftel/publications/about_oftel/index.htm.

2. Data from ww.crtc.gc.ca/eng/BACKGRND/Brochures/B19903.htm (November 26, 2002).

3. Data from http://europa.eu.int/information_society/topics/telecoms/regulatory/index_en.htm (November 26, 2002).

4. See the CRTC Telecom Decision 1997–08 (www.crtc.gc.ca/archive/eng/Decisions/1997/DT97-8.htm).

5. CRTC, Telecom Decision 2002–34.

6. AT&T Canada began as Unitel, a joint venture of two Canadian companies (one of which was the Canadian Pacific Railroad). Unitel later entered into an agreement with AT&T, and when the Canadian investors pulled out, they left the failing company in AT&T's hands. AT&T was subsequently induced to write off several billion dollars in debt as it withdrew. The company then became Allstream, which was purchased by Manitoba

Telecom in 2004, becoming MTS-Allstream. See www.crtc.gc.ca/eng/publications/reports/telecom_summary.htm.

7. In 2001 long-distance revenues in Canada (including international) totaled Can$6,105 million, and long-distance minutes 40,500 million (www.crtc.gc.ca/eng/publications/reports/telecom_summary.htm). At an exchange rate of Can$1.55 to US$1, the average price per minute was US$0.097. The comparable estimate for the United States was $0.10 per minute in 2001, according to the FCC's *Telecommunications Industry Revenues, 2002* (March 2004), table 9 (www.fcc.gov/Bureaus/Common_Carrier/Reports/FCC-State_Link/IAD/telrev00.pdf). See also Robert W. Crandall and Thomas W. Hazlett, "Telecommunications Policy Reform in the United States and Canada," in *Telecommunications Liberalization on Two Sides of the Atlantic*, edited by Martin Cave and Robert W. Crandall (AEI-Brookings Joint Center for Regulatory Studies, 2001).

8. Industry Canada, *Telecommunications Services in Canada: An Industry Overview* (Ottawa, October 31, 2002).

9. This is the U.S. equivalent of Chapter 11 bankruptcy.

10. "At September 30, 2002, the liability of $3.5 billion was reflected as 'AT&T Canada obligation.' On October 8, 2002, Tricap Investments Corporation, a wholly owned subsidiary of Brascan Financial Corporation, purchased an approximate 63% equity interest in AT&T Canada and CIBC Capital Partners acquired an approximate 6% equity interest in AT&T Canada. AT&T paid the purchase price for the AT&T Canada shares on behalf of Tricap and CIBC Capital Partners. AT&T funded the purchase price of the AT&T Canada shares partly with the net proceeds of approximately $2.5 billion received from the sale of 230 million shares of AT&T common stock on June 11, 2002. The remaining portion of the obligation was financed through short-term sources. Tricap and CIBC Partners made a nominal payment to AT&T upon completion of the transaction. AT&T continues to hold a 31% ownership interest in AT&T Canada." Quarterly Report to the Securities and Exchange Commission, Form 10-Q (November 13, 2002).

11. See "360 Networks to Pay C$260.5 Million for Group Telecom," *Reuters,* November 20, 2002.

12. BCE sold the eastern Canadian assets to Call-Net in order to obtain regulatory approval.

13. There are five principal incumbents: Bell Canada, Aliant, TELUS, MTS, and Sasktel. However, Sasktel is owned by the province of Saskatchewan. The other four companies had $28 billion in market capitalization (June 30, 2004) and $16 billion in book value of long-term debt. Data from company reports and Investcom (www.investcom.com/cgi-bin/nameindustry/industrysearch.cgi).

14. OECD data (www.investcom.com [October 2003]).

15. Ibid.

16. These directives may be found at http://europa.eu.int/information_society/topics/telecoms/regulatory/98_regpack/index_en.htm.

17. European Union, Regulation 2887/2000 of the European Parliament and of the Council on Unbundled Access to the Local Loop, December 7, 2000.

18. European Commission, *Ninth Report on the Implementation of the Telecommunications Regulatory Package* (Brussels, 2003), annex 1, fig. 14.

19. Data from European Competitive Telecommunications Association (ECTA), *Scorecard* (December 2003) (www.ectaportal.com).

20. Ibid.

21. Ibid.

22. Ibid.

23. Data from the company's website, www.tele2.com/index.html.

24. Japan Ministry of Public Management, Home Affairs, Posts and Telecommunications, *Information and Communications in Japan,* White Paper 2002 (Tokyo, 2002), p. 47. The market share data are for calls, not revenues.

25. These data are from a presentation by Yasuhiko Taniwaki, counselor, Embassy of Japan, at the Center for International and Stategic Studies, Washington, 2004, based on data supplied by the Ministry of Public Management, Home Affairs, Posts and Telecommunications. See also Matsasugu Tsuji, "Network Competition in the Japanese Broadband Infrastructure: From the Last One Mile to the Last Quarter Mile," and Hidenori Fuke, "Facilities-Based Competition and Broadband Access," papers presented at the International Conference on Convergence in Communications Industries, Warwick Business School, University of Warwick, November 4–5, 2002.

26. The incumbent, NTT, is hindered by a regulation that forbids it to offer a broadband service integrated with its own ISP service. NTT now appears to be planning on widespread deployment of direct fiber to the end user, which would offer subscribers much greater speed than is available from DSL.

27. Softbank Press Release, July 9, 2004 (www.softbank.co.jp/english/index.html).

28. Operating losses in fiscal 2001 equaled 85 percent of revenues. See Fuke, "Facilities-Based Competition and Broadband Access."

29. All data are from the Softbank Annual Report for Fiscal 2004 (www.softbank.co.jp/english/index.html).

30. Ibid.

31. Softbank Press Release, July 9, 2004.

32. See Kazuhiro Shimamura, "A Phone Challenge in Japan," *Wall Street Journal*, September 9, 2004, p. C-4.

33. Junseong An, "E-Korean DSL Policy: Implications for the United States," *John Marshall Journal of Computer & Information Law* (Spring 2002): 421.

34. Ibid.

35. Korea Ministry of Information and Communication, *Data Bank* (www.mic.go.kr/index.jsp).

36. For details on each company's extensive network, see Korea Ministry of Information and Communication, *White Paper, Internet Korea 2004* (www.mic.go.kr/index.jsp), chap. 5.

37. Annual reports of the companies.

38. Korea Ministry of Information and Communication, *IT Korea 2002* (www.mic.go.kr/index.jsp), p. 24.

39. Korea Ministry of Information and Communication, *White Paper, Internet Korea 2004.*

40. The conventional wisdom is that Koreans have a much stronger interest in electronic games and that broadband permits subscribers to engage in real-time, interactive games from their residences. The annual revenues of the online game business recently reached 298.5 billion won, or about $250 million, which is equal to about $5 per person per year. See Ministry of Information and Communication, *White Paper: Internet Korea 2003.*

41. Bureau of Labor Statistics, *Consumer Price Index, All Urban Households*, "Land-line intrastate toll calls," accessible at www.bls.gov/cpi/home.htm (February 2005).

42. European Commission, *Ninth Report on the Implementation of the Telecommunications Regulatory Package,* annex 1. The average price of a local three-minute call rose from 12.5 Euro-cents in August 1998 to 13.5 Euro-cents in August 2001. The average price of a ten-minute local call declined from 39.8 Euro-cents to 38.9 Euro-cents over the same period.

43. Alessandra Colecchia and Paul Schreyer, *ICT Investment and Economic Growth in the 1990s: Is the United States a Unique Case? A Comparative Study of Nine OECD Countries,* STI Working Paper 2001/7 (OECD Directorate for Science, Technology and Industry, October 25, 2001).

44. Data from the Bureau of Economic Analysis, National Income and Output Accounts.

45. The bars in figure 9-4 show the value of a share of common stock of each company in February 2000 and June 2004 compared with December 1997, just before the European Union liberalized the sector. Because several of the companies were privatized in the late 1990s, it is impossible to go back to a period that clearly antedates the stock market "bubble."

Notes to Chapter Ten

1. For the views of a former regulator and economist on this phenomenon, see Alfred E. Kahn, *Letting Go: Deregulating the Process of Deregulation* (Michigan State University Institute of Public Utilities and Network Industries, 1998).

2. National Cable and Telecommunications Association, "2004 Mid-Year Industry Overview" (www.ncta.com [2004]).

3. For a contrary view, see Bruce M. Owen, *The Internet Challenge to Television* (Harvard University Press, 2000).

4. Cable subscriptions grew by nearly 15 percent between 1996 and 2002 while wireless subscribers more than tripled. These growth rates are not sustainable because nearly 90 percent of all U.S. households now subscribe to cable or satellite services and half of all U.S. residents now have cell phones.

5. Between 2000 and 2003, end-user telecom revenues for fixed-wire carriers declined from $171.7 billion to $149.2 billion. See the FCC's periodic *Telecommunications Industry Revenue Data* reports.

6. Market value of equity plus book value of long-term obligations.

7. As this book goes to press, the FCC has issued a *Notice of Proposed Rulemaking* on the regulatory issues raised by VoIP, but it has not reached any conclusions on the major issues. See "In the Matter of IP-enabled Services," WC Docket 04-36 (March 10, 2004). In this notice, the commission asks ominously for comments on how "the regulatory classification of IP-enabled services, including VoIP, would affect the Commission's ability to fund universal service." Surely, it already knows the answer to this question.

8. See Clifford Winston, "Economic Deregulation: Days of Reckoning for Microeconomists," *Journal of Economic Literature* 31 (September 2003): 1263–89.

9. For a discussion of this example of regulatory protection, see Morris A. Adelman, *A&P: A Study in Price-Cost Behavior and Public Policy* (Harvard University Press, 1959).

10. For an interesting discussion of many of these exercises in regulatory protection, see Roger G. Noll and Bruce M. Owen, eds., *The Political Economy of Deregulation* (Washington: American Enterprise Institute, 1983).

11. A complete tabulation of these Universal Service disbursements may be found in the FCC's *Trends in Telephone Service*, annual editions.

12. FCC, *Statistics of Communications Common Carriers*, annual editions.

13. This proposal surfaced in August 2004, and therefore its prospects are far from clear as this book goes to press. See, for example, Anna-Maria Kovacs and Kristin L. Burns, *TELECOM NOTE: Intercarrier Compensation Proposal* (Regulatory Source Associates LLC, August 14, 2004).

14. Verizon and Bell South have not supported the proposal, in part because they apparently fear that they cannot pass on the increased fixed monthly rates to their subscribers.

15. Robert W. Crandall and Leonard Waverman, *Who Pays for Universal Service? When Telephone Subsidies Become Transparent* (Brookings, 2000), chap. 6.

16. FCC, *Proposed Fourth Quarter 2003 Universal Service Contribution Factor*, CC Docket 96-45 (September 5, 2003).

17. Crandall and Waverman, *Who Pays for Universal Service?* p. 162; Jerry A. Hausman, "Taxation by Telecommunications Regulation," *Tax Policy and the Economy* 12 (1998): 29–48.

18. Robert Litan and Roger Noll, "The Uncertain Future of the Telecommunications Industry," paper presented at the Brookings Instution forum *Turmoil in Telecom,* December 2, 2003.

19. The examples provided were drawn from telephony, motor vehicles, motion pictures, commercial television, and cable television.

20. See Bruce M. Owen and Gregory L. Rosston, "Local Broadband Access: *Primum Non Nocere or Primum Processi*?" (Stanford Institute for Economic Policy Research, July 2003). For the opposite view from a lawyer, see Lawrence Lessig, *The Government's Role in Promoting the Future of Telecommunications Industry Broadband Deployment,* Hearings before the U.S. Senate Commerce Committee, 107 Cong. 2 sess. (Government Printing Office, 2002).

Notes to Appendix

1. The Reg variables are dummy variables reflecting the nature of regulation in each state. In addition, I use dummy variables for each Bell company to capture any company-specific effects.

2. The detailed results for all regressions are available from the author.

3. Anna Maria Kovacs and Kristin Burns, "The Status of 271 and UNE Platform in the Regional Bells' Territories" (Commerce Capital Markets, April 2002); Billy Jack Gregg, *A Survey of Unbundled Network Element Prices in the United States* (National Regulatory Research Institute), periodic issues; Kevin A. Hassett, Zoya Ivanova, and Laurence J. Kotlikoff, "Increased Investment, Lower Prices—The Fruits of Past and Future Telecom Competition," paper presented at the American Enterprise Institute, Washington, September 2003.

4. CTIA, *Semi-Annual Wireless Industry Survey*, downloaded from www.wow-com.com/industry/stats/surveys/, and NCTA, *Industry Overview*, downloaded from www.ncta.com/docs/pagecontent.cfm?pageID=86.

5. In addition to the regression results reported in table 5-7, I estimated twelve separate regressions of *annual* RBOC capital expenditures in each of four years—2000, 2001, 2002, and 2003—on the exogenous variables shown in table A-1, excluding UNE-P but using the

FCC data on CLEC share for each of three years, 2001, 2002, and 2003. I estimated another twelve separate regressions for only those states (thirty-four) for which the FCC reported CLEC share data in every year for 2001–03. Thus twenty-four regressions were estimated in toto. The coefficients for the CLEC share variable were positive in nineteen cases and negative in five cases. Of the nineteen positive coefficients, only two were statistically significant at the 5 percent confidence level and another three at the 10 percent level.

Index

THE BROOKINGS INSTITUTION

The Brookings Institution is an independent organization devoted to nonpartisan research, education, and publication in economics, government, foreign policy, and the social sciences generally. Its principal purposes are to aid in the development of sound public policies and to promote public understanding of issues of national importance. The Institution was founded on December 8, 1927, to merge the activities of the Institute for Government Research, founded in 1916, the Institute of Economics, founded in 1922, and the Robert Brookings Graduate School of Economics and Government, founded in 1924. The Institution maintains a position of neutrality on issues of public policy to safeguard the intellectual freedom of the staff. Interpretations or conclusions in Brookings publications should be understood to be solely those of the authors.